Tom Adams, Steve Langfield and Averil Macdonald

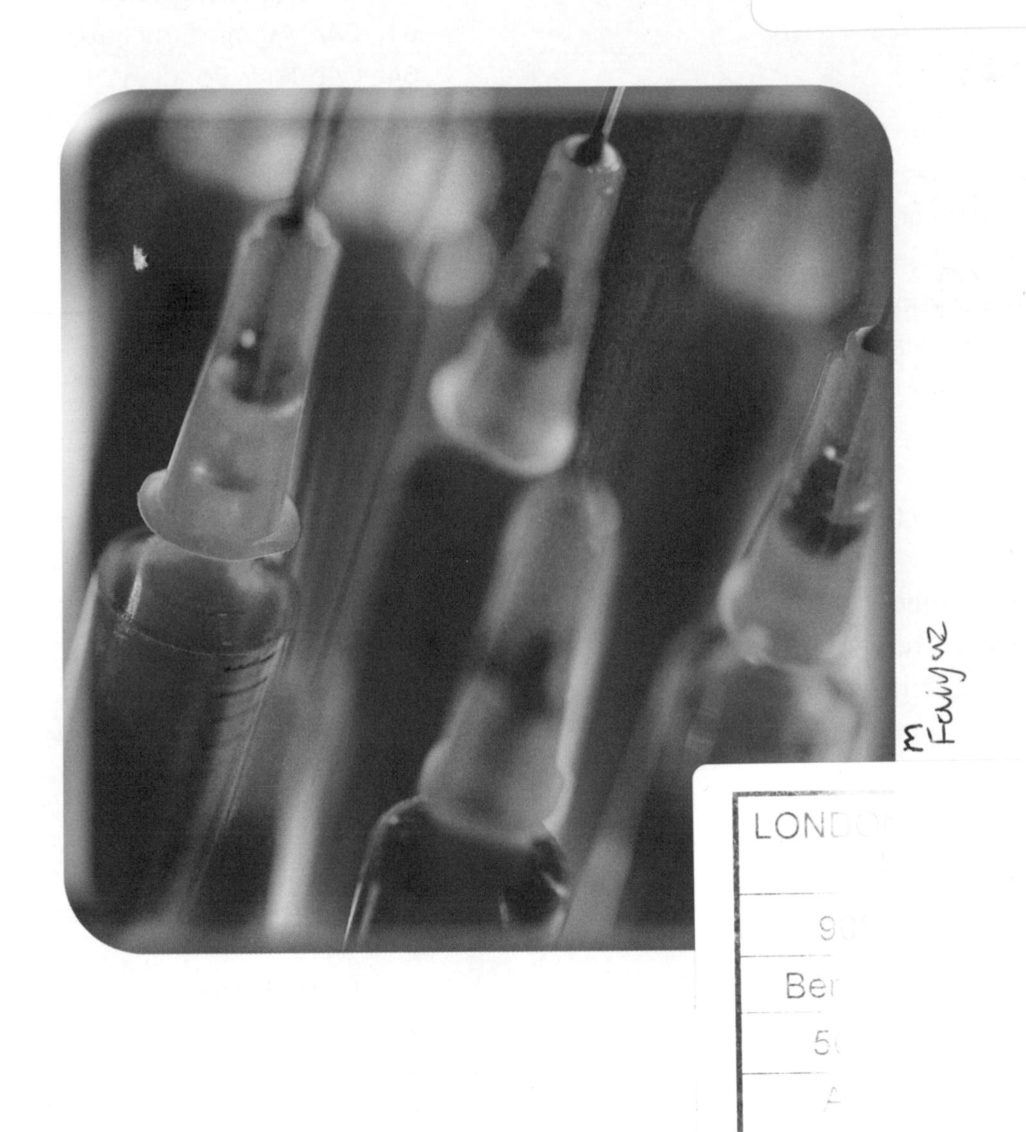

LONDO
90
Ber
5
A

# REVISION PLUS

## OCR Gateway
## GCSE Additional Science B
Revis... Companion

NEWHAM LIBRARIES
9080010011679

# Contents


1 Fundamental Scientific Processes

## Biology

**B3: Living and Growing**
3 B3a Molecules of Life
5 B3b Proteins and Mutations
7 B3c Respiration
9 B3d Cell Division
12 B3e The Circulatory System
14 B3f Growth and Development
16 B3g New Genes for Old
18 B3h Cloning
21 B3 Exam Practice Questions

**B4: It's a Green World**
22 B4a Ecology in the Local Environment
25 B4b Photosynthesis
28 B4c Leaves and Photosynthesis
30 B4d Diffusion and Osmosis
33 B4e Transport in Plants
36 B4f Plants Need Minerals
37 B4g Decay
39 B4h Farming
42 B4 Exam Practice Questions

## Chemistry

43 Fundamental Chemical Concepts

**C3: Chemical Economics**
47 C3a Rate of Reaction (1)
49 C3b Rate of Reaction (2)
51 C3c Rate of Reaction (3)
52 C3d Reacting Masses
54 C3e Percentage Yield and Atom Economy
55 C3f Energy
57 C3g Batch or Continuous?
58 C3h Allotropes of Carbon and Nanochemistry
60 C3 Exam Practice Questions

**C4: The Periodic Table**
61 C4a Atomic Structure
64 C4b Ionic Bonding
67 C4c The Periodic Table and Covalent Bonding
69 C4d The Group 1 Elements
71 C4e The Group 7 Elements
73 C4f Transition Elements
74 C4g Metal Structure and Properties
75 C4h Purifying and Testing Water
77 C4 Exam Practice Questions

## Physics

**P3: Forces for Transport**
78 P3a Speed
80 P3b Changing Speed
83 P3c Forces and Motion
85 P3d Work and Power
87 P3e Energy on the Move
89 P3f Crumple Zones
91 P3g Falling Safely
92 P3h The Energy of Games and Theme Rides
94 P3 Exam Practice Questions

**P4: Radiation for Life**
95 P4a Sparks
97 P4b Uses of Electrostatics
98 P4c Safe Electricals
101 P4d Ultrasound
103 P4e What is Radioactivity?
105 P4f Uses of Radioisotopes
106 P4g Treatment
107 P4h Fission and Fusion
110 P4 Exam Practice Questions

111 Answers
115 Glossary
118 Periodic Table
IBC Index

Scientists carry out **investigations** and collect **evidence** in order to explain how and why things happen. Scientific knowledge and understanding can lead to the **development of new technologies** that have a huge impact on **society** and the **environment**.

**Scientific evidence** is often based on data collected through **observations** and **measurements.** To allow scientists to reach conclusions, evidence must be **repeatable, reproducible** and **valid.**

## Models

**Models** are used to explain scientific ideas and the Universe around us. Models can be used to describe:

- a complex idea – like how heat moves through a metal
- a system – like the Earth's structure.

Models make systems or ideas easier to understand by including only the most important parts. They can be used to explain real-world observations or to make predictions. But, because models don't contain all the **variables**, they sometimes make incorrect predictions.

Models and scientific ideas may change as new observations are made and new **data** are collected. Data and observations may be collected from a series of experiments. For example, the accepted model of the structure of the atom has been modified as new evidence has been collected from many experiments.

## Hypotheses

Scientific explanations are called **hypotheses** – these are used to explain observations. A hypothesis can be tested by planning experiments and collecting data and evidence. For example, if you pull a metal wire you may observe that it stretches. This can be explained by the scientific idea that the atoms in the metal are arranged in layers that can slide over each other. A hypothesis can be modified as new data is collected, and may even be disproved.

## Data

**Data** can be displayed in **tables**, **pie charts** or **line graphs.** In your exam you may be asked to:

- choose the most appropriate method for displaying data
- identify trends
- use data mathematically – including using statistical methods, calculating the mean and calculating gradients of graphs.

**A Table**

| % Yield / Pressure | Temperature | | | |
|---|---|---|---|---|
| | 250°C | 350°C | 450°C | 550°C |
| 200 atm | 73% | 50% | 28% | 13% |
| 400 atm | 77% | 65% | 45% | 26% |

**A Pie Chart**

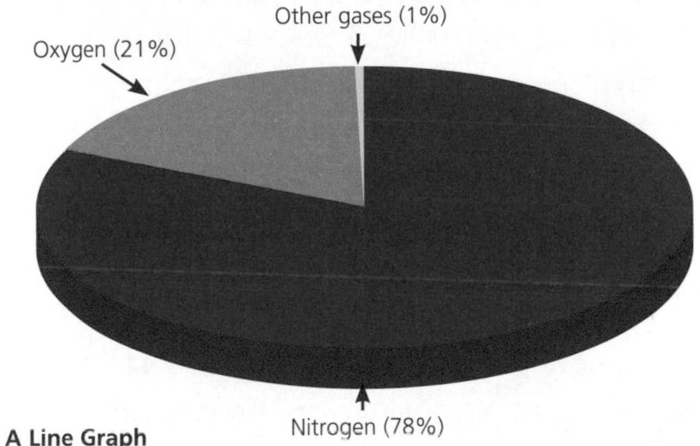


**A Line Graph**

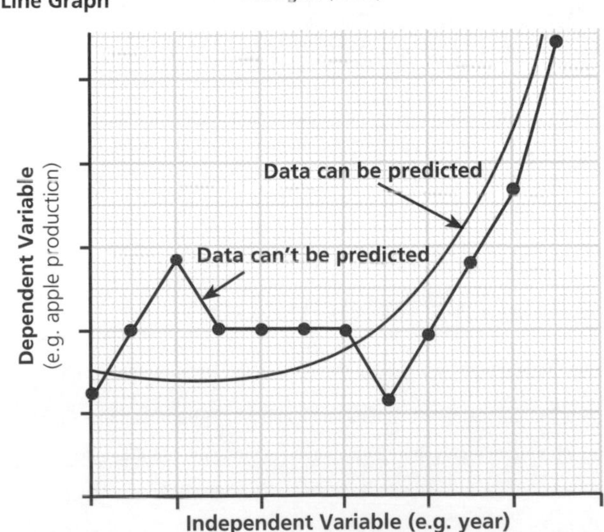

## Data (cont)

Sometimes the same data can lead to different conclusions. For example, data shows that the world's average temperatures have been rising significantly over the last 200 years. Some scientists think this is due to increased combustion of fossil fuels, while other scientists think it's a natural change that has happened before during Earth's history.

## Scientific and Technological Development

Every scientific or technological development can have effects that we do not know about. This can give rise to **issues**. An issue is an important question that is in dispute and needs to be settled. Issues can be:

- **social** – they impact on the human population of a community, city, country or the world
- **environmental** – they impact on the planet, its natural ecosystems and resources
- **economic** – money and related factors such as employment and the distribution of resources
- **ethical** – what is right and wrong morally; a value judgement must be made
- **cultural** – giving an insight into differences between people on local and global scales.

**Peer review** is a process of self-regulation involving experts in a particular field who **critically examine** the work undertaken. Peer review methods are designed to maintain standards and provide **credibility** for the work that has been carried out. The methods used vary depending on the nature of the work and also on the overall purpose behind the review process.

## Evaluating Information

**Conclusions** can then be made based on the scientific evidence that has been collected – they should try to explain the results and observations.

**Evaluations** look at the whole investigation. It is important to be able to evaluate information relating to social–scientific issues. **When evaluating information:**

- make a list of **pluses** (pros)
- make a list of **minuses** (cons)
- consider how each point might **impact on society**.

You also need to consider if the source of information is reliable and credible and to consider opinions, bias and weight of evidence.

**Opinions** are personal viewpoints – those backed up by valid and reliable evidence carry far more weight than those based on non-scientific ideas. Opinions of experts can also carry more weight than those of non-experts. Information is **biased** if it favours one particular viewpoint without providing a balanced account. Biased information might include incomplete evidence or it might try to influence how you interpret the evidence.

Examples of these processes are included within the main content of the book. However, it is important to remember that fundamental scientific processes are relevant to all areas of science.

## B3: Living and Growing

This module looks at:

- DNA, how its structure was discovered and how it provides the template for proteins.
- The structure and function of proteins, how enzymes work and how mutations arise.
- Respiration and how energy is used in metabolic processes.
- Cell division for growth, and the production of gametes.
- The heart and circulatory system, blood and blood disorders.
- How animals and plants grow and develop, the use of stem cells and related ethical issues.
- Genetic engineering, genetic modification and selective breeding.
- Modern cloning technology, and producing cuttings and tissue cultures in plants.

## Cells

The fundamental processes of life take place inside **cells**. Below is a typical animal cell:

**A Cheek Cell from a Human**

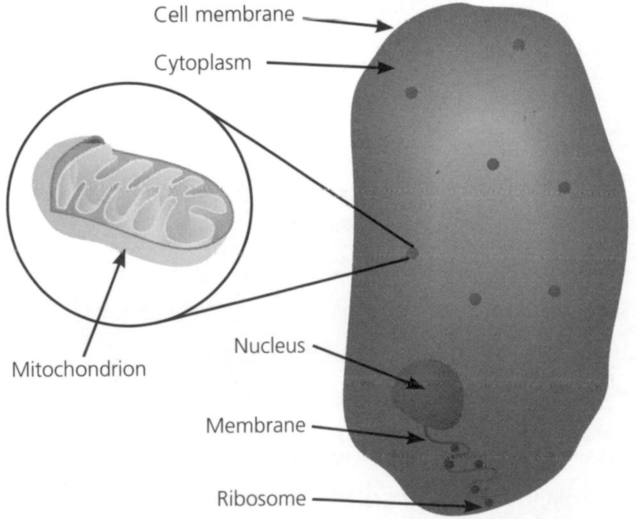


- Most chemical reactions take place in the **cytoplasm**. It may contain mitochondria which is where most energy is released in respiration.
- The **cell membrane** controls movement into and out of the cell.
- The **nucleus** contains the genetic information and controls what the cell does. It has a membrane extending from it, onto which ribosomes are attached.
- Respiration takes place in the **mitochondria**, providing energy for life processes. Liver and muscle cells have particularly large numbers of mitochondria because they have a high energy requirement.

## DNA

The nucleus of each cell contains a complete set of genetic instructions called the **genetic code**. The information is carried by **genes** on **chromosomes**, which are long coils of a chemical called **DNA** (deoxyribonucleic acid). The genetic code controls cell activity and, consequently, some characteristics of the whole organism.

A DNA molecule is made of two strands coiled around each other in a **double helix** (spiral). The genetic code is in the form of a chemical code made up of four **bases**. These bases bond together in pairs, forming the cross-links (like rungs on a ladder) which hold the two strands of DNA together.

This highly specialised structure was first put forward in 1953 by scientists **James Watson** and **Francis Crick**. They used data from X-ray crystallography experiments performed by Rosalind Franklin. The data from her experiments helped Watson and Crick develop a model that fitted the data. The model showed two chains wound as a double-helix and other data indicated that the bases occurred in pairs.

Each gene in a DNA molecule contains a different sequence of bases. The bases code for the structure of different proteins that are needed for the growth and repair of cells. Proteins are made in the cytoplasm. The code, carried in the DNA, is in the nucleus. As DNA does not leave the nucleus, a copy needs to be made and transferred to the cytoplasm.

HT DNA controls cell function by controlling the production of proteins, some of which are enzymes.

## DNA (cont)

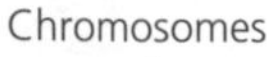


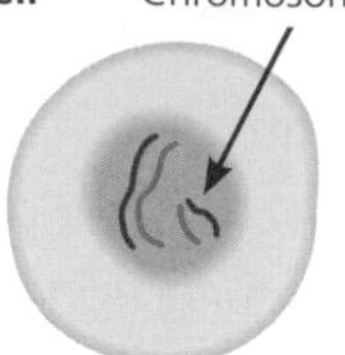

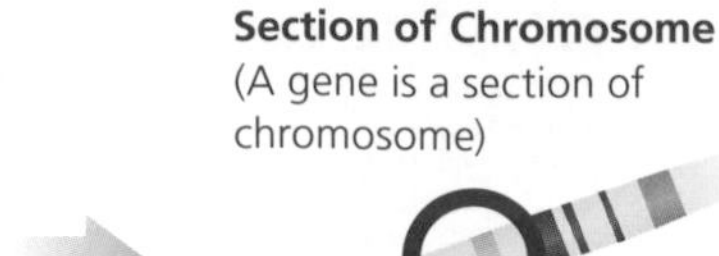

### Bases in DNA

The four bases in DNA are A, C, G and T. A always bonds with T, and C always bonds with G on opposite strands of the DNA molecule. This is **complimentary base pairing**.

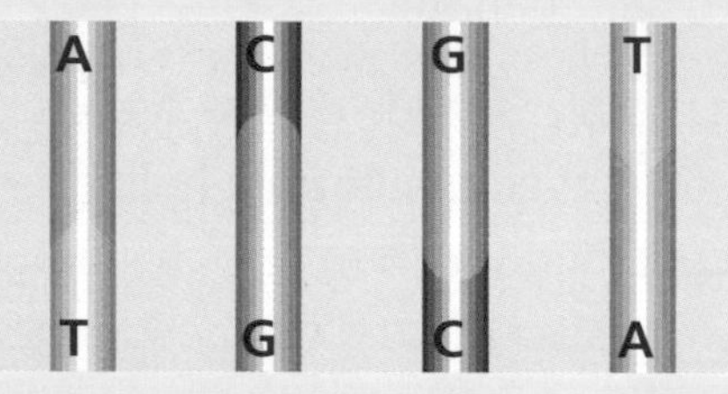


Proteins are made up of chains of **amino acids**. The cell uses the amino acids from food to construct them. When a cell needs to produce proteins of a particular type, certain genes will be 'switched on'. These genes and their resulting proteins determine the characteristics of cells.

### Protein Synthesis

The sequence of bases in a gene represents the order in which the cell should assemble amino acids to make the protein. A group of **three** bases represents **one** amino acid in a protein chain. Each protein has a different shape and function.

Proteins are synthesised by structures called **ribosomes**. These are too small to be seen under the light microscope and have to be viewed under an electron microscope. They are located in the cytoplasm.

In order for the DNA code in the nucleus to be translated as new protein by the ribosomes, a 'messenger' molecule needs to travel from the nucleus into the cytoplasm and then become attached to a ribosome. The molecule is called **mRNA**.

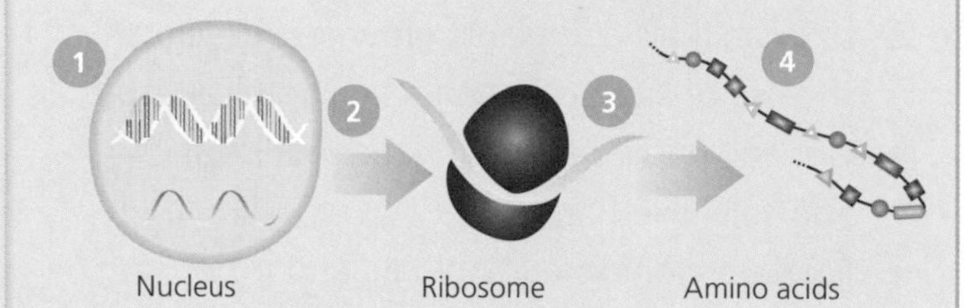


1. mRNA constructed from DNA template in nucleus.
2. mRNA carries code from original DNA into the cytoplasm and becomes attached to a ribosome.
3. Ribosome 'translates' the code on mRNA and constructs a sequence of amino acids.
4. Amino acids linked together as a long chain protein, which is released from the ribosome to carry out its function.

### More on Watson and Crick

Discoveries such as that by Watson and Crick illustrate how scientists have to work together and often develop theories based on work that may have been done many years before. The structure so perfectly fitted the experimental data that it was accepted almost immediately. This is not often the case in science where findings from experiments have to be tested again by other scientists to ensure that conclusions are valid and accurate.

Watson and Crick were awarded the Nobel Prize for Medicine in 1962 for their discovery of the structure of DNA. Their discovery is considered the most important biological work of the last century. It led to an explosion of scientific advances including genetic engineering and mapping the human genome.

## Proteins

Proteins are needed in cells because they are a vital component in the construction of membranes. Some proteins are very specialised for particular functions:

- **Collagen** is an important structural protein found in connective tissue.
- **Insulin** is a hormone that helps to control blood sugar levels.
- **Haemoglobin** is a carrier protein, found in red blood cells, which binds with oxygen.
- **Enzymes** control chemical reactions in the body.

HT Each protein has its own number and sequence of amino acids. This results in different shapes of molecule. It is estimated that there are over 19 000 different proteins in the human body, each with a particular function.

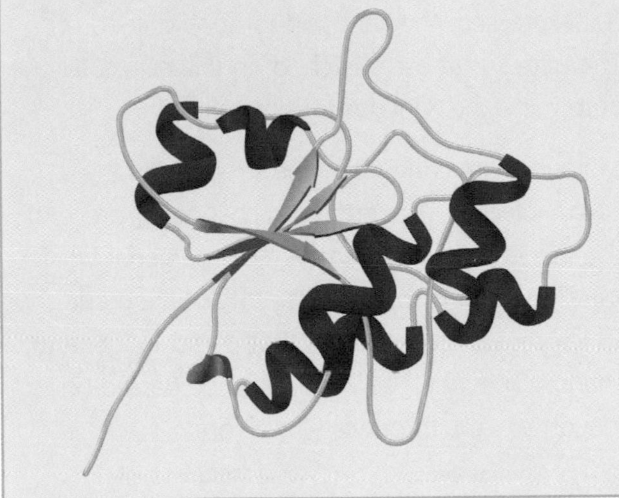

## Enzymes

Enzymes are proteins that act as **biological catalysts.** They speed up chemical reactions, including those that take place in living cells, e.g. respiration, photosynthesis and protein synthesis. Enzymes are highly **specific**; each one will speed up only a particular reaction. Enzymes have active sites, regions of the enzyme molecule which bind to substrate molecules temporarily and allow the substrate to be changed.

Changing temperature and pH will affect the rate of a reaction catalysed by an enzyme.

## Enzyme Activity and Temperature

The graph below shows the effect of temperature on enzyme activity:

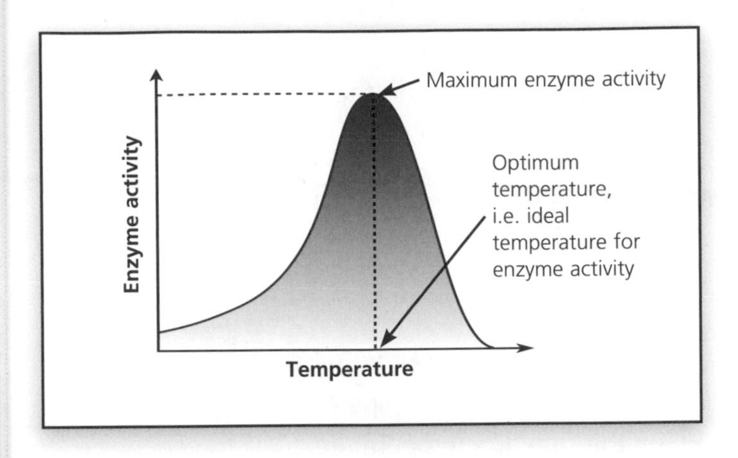


A rise in temperature increases the number of collisions between reactants and enzymes, and will increase the enzyme activity until the optimum temperature is reached. Temperatures above the optimum permanently damage the enzyme molecules, decreasing or stopping enzyme activity.

Different enzymes have different optimum temperatures. The ones in the human body work best at about 37°C.

## Enzyme Activity and pH

The graph below shows how changes in pH affect enzyme activity.

There is an optimum pH at which the enzyme works best. As the pH increases or decreases, the enzyme becomes less and less effective.

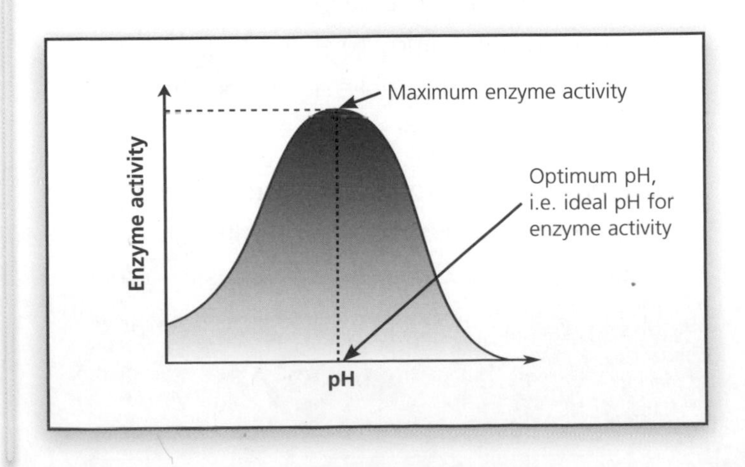

# B3 Proteins and Mutations

## Enzyme Activity and pH (cont)

The optimum pH for different enzymes can vary considerably. For example:

- the enzyme in human saliva works best at about pH 7.3
- the enzyme pepsin (in the stomach) needs very acidic conditions in order to work well.

### HT Measuring the Rate of a Reaction

The rate of an enzyme-controlled reaction can be expressed as a **$Q_{10}$ value** by comparing the rate at a particular temperature and dividing it by the rate at a temperature 10°C lower:

$$Q_{10} = \frac{\text{Rate at temperature, } t}{\text{Rate at temperature, } t - 10°\text{C}}$$

## The Lock and Key Mechanism

Each enzyme has a different number and sequence of amino acids. This gives it a unique 3-D shape, which includes an active site that only a specific reactant can fit into (like a key in a lock).

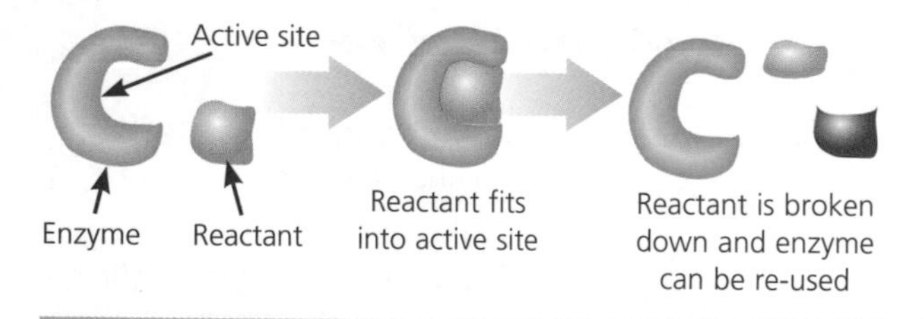


HT Enzyme molecules are **denatured** by high temperature and extreme pH; the bonds in the protein break and the shape of its active site is changed irreversibly, so the lock and key mechanism no longer works.

At lower than optimum temperatures, the enzyme does not collide as often with substrate molecules and so the reaction rate is lowered.

## Mutations

Different cells and different organisms produce different proteins. If the base sequence changes, this might result in the production of different proteins, in which cases new forms of genes may arise from mutations (changes) in existing genes.

Mutations are usually **harmful** or have no effect whatsoever but can sometimes be beneficial because the new protein could function better. They occur naturally and spontaneously but their frequency is increased by exposure to:

- ultraviolet light
- radioactive substances
- X-rays
- certain chemicals.

HT Every cell contains a complete set of genes to code for the whole body. But only some of these genes are used in any one cell. Some genes are not expressed; they are said to be 'switched off'. The genes that are switched on will eventually determine the function of the cell.

Mutations are changes to the structure of the DNA molecule. The mutations change (or prevent) the sequence of amino acids that the genes usually code for, so different proteins are made, which causes the nature of that particular gene to change. The 'new' gene can then be passed on to 'daughter' cells through cell division.

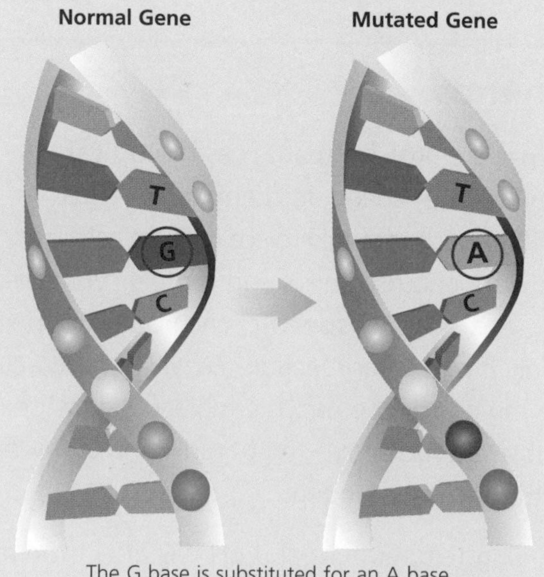


The G base is substituted for an A base

## Providing Oxygen to Cells

Living cells need energy to function and during exercise they need much more energy. Examples of the body's energy requirements include muscle contraction, protein synthesis and control of body temperature in mammals. It must not be forgotten that plants respire as well. They need energy for making new substances, growth and transport of materials.

### HT Measuring Energy Release

The sum of the chemical reactions of the body is called the **metabolic rate**. It is useful for scientists to be able to measure this rate in experiments so that they can investigate factors that might affect it. Since aerobic respiration requires oxygen, a good measure is how quickly an organism, tissue or cell can take up oxygen; this is called the **oxygen consumption rate**. Carbon dioxide output can also be used as a measure of metabolic rate. In practice, this will always be an estimate because living tissue often uses both aerobic and anaerobic respiration depending on circumstances. Also, due to the fact that respiration is a series of enzyme-controlled reactions, the rate will be influenced by temperature and pH.

The circulatory system carries **oxygen** and **glucose** to all the body's cells so that **energy** can be released through **aerobic respiration**.

Aerobic respiration takes place inside the cells. The oxygen and glucose molecules react and the glucose molecules are broken down to release energy. It involves a combination of chemical reactions which can be simplified as a single equation:

| glucose + oxygen ⟶ carbon dioxide + water | Energy released |
|---|---|
| $C_6H_{12}O_6 + 6O_2 \longrightarrow 6CO_2 + 6H_2O$ | Energy released |

HT Energy from respiration is locked up in a molecule called **ATP**. ATP is called upon when the cell requires energy to be released.

**A Working Muscle Cell**

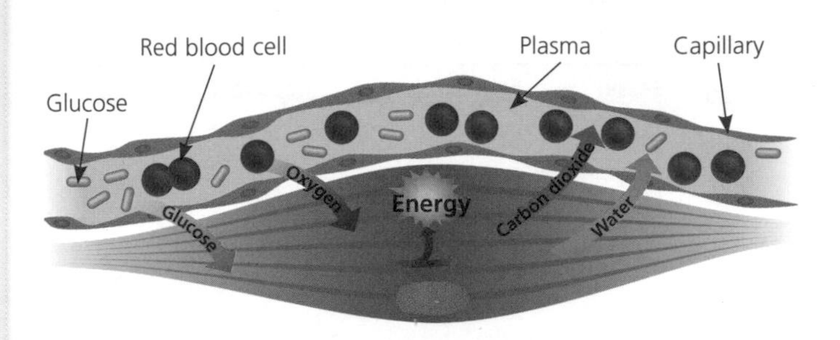


You may be asked to analyse data to compare respiration rates to include increased oxygen consumption and increased carbon dioxide production.

## Respiratory Quotient

The **respiratory quotient** measures the ratio between the oxygen that an organism takes in and the carbon dioxide that it gives out. To determine which type of substrate (molecule) is being respired in an organism, a respiratory quotient can be calculated using the formula:

$$RQ = \frac{\text{Carbon dioxide produced}}{\text{Oxygen used}}$$

- The respiration of carbohydrate results in an RQ of 1.0.
- The respiration of fat gives an RQ of 0.7.
- The respiration of protein gives an RQ of 0.8.

## Effect of Exercise

When a person exercises, their breathing and pulse rates increase to deliver oxygen and glucose to the cells in their muscles more quickly. This increase helps to **remove** carbon dioxide produced during respiration more quickly.

## Effect of Exercise (cont)

To investigate how heart rate responds to exercise, measure your resting rate immediately after exercise. Then measure it every minute until it returns to normal – the time taken to do this is called your **recovery rate**.

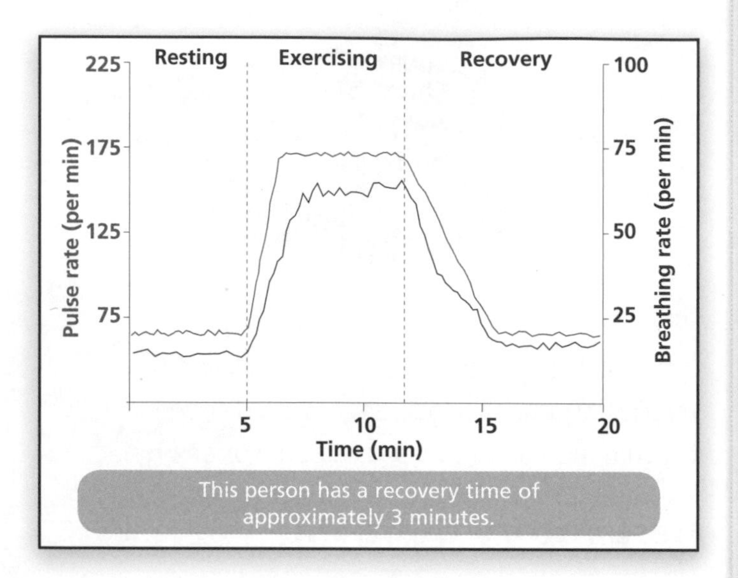


This person has a recovery time of approximately 3 minutes.

## Anaerobic Respiration

Anaerobic respiration takes place in the **absence of oxygen**. It quickly releases a small amount of energy per gram of glucose through the **incomplete** breakdown of glucose.

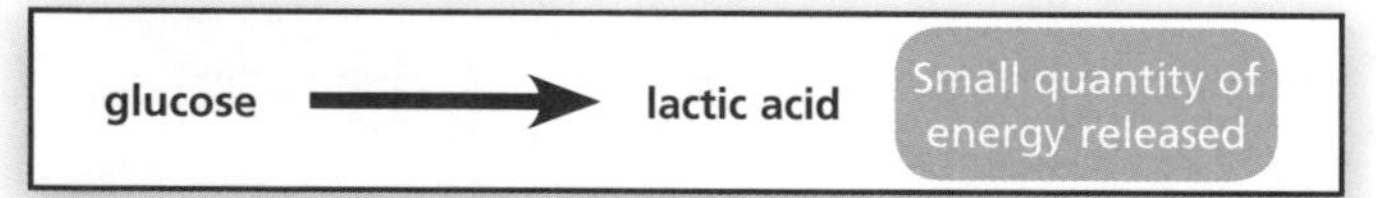


Anaerobic respiration occurs when the muscles are working so hard that the lungs and circulatory system cannot deliver enough **oxygen** to break down all the available glucose through aerobic respiration.

At this point, anaerobic respiration starts to take place **in addition** to aerobic respiration.

The **lactic acid** produced during anaerobic respiration is relatively toxic to the cells and can cause pain (cramp) and a sensation of **fatigue** in the muscles.

Because anaerobic respiration involves the incomplete breakdown of glucose, much less energy (about a twentieth) is released than in aerobic respiration. However, it can produce energy much faster over a short period of time, until fatigue sets in. This makes anaerobic respiration a real necessity in events that require a short, intense burst of energy, e.g. the 100-metre sprint.

### HT Recovering After Anaerobic Respiration

Lactic acid produced when intense exercise is undertaken must be broken down fairly quickly and removed to avoid cell damage and to relieve the feeling of fatigue.

Immediately after anaerobic exercise:

- **the heart rate stays high** – pumping blood through the muscles to remove the lactic acid and transport it to the liver to be broken down
- **deep breathing or 'panting' continues** – ensuring enough oxygen is taken in to oxidise the lactic acid (producing carbon dioxide and water).

In effect, the body is taking in the oxygen that was not available for aerobic respiration during exertion. This is why the process is sometimes referred to as repaying the **oxygen debt**.

## Single-cell and Multi-cell Organisms

Single-cell organisms (e.g. amoeba) are very small and have to rely on diffusion to obtain glucose and oxygen, and to remove waste products.

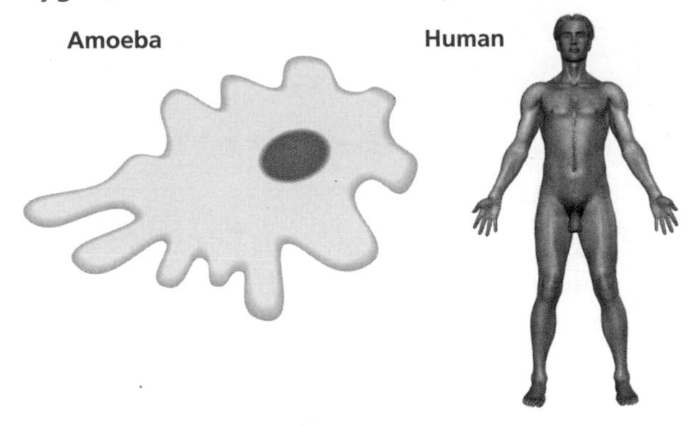


Multi-cell organisms (e.g. humans) are much larger and more complex. This means that they require specialised organs to carry out functions like gas exchange and digestion. When cells develop a specialised structure to carry out a specific function, this is called **differentiation**.

HT Specialised cells build up into specialised tissues and organs. These, in turn, form specialised systems, which are needed for:
- communication between cells
- supplying the cells with nutrients
- controlling exchange of materials with the environment, e.g. breathing.

## Cell Division

In mammals, body cells are **diploid**, which means they contain two sets of matching chromosomes.

### Mitosis

**Mitosis** is the process by which a diploid cell divides to produce two more diploid cells. It produces new cells:
- for **growth**
- to **repair** damaged tissue
- to **replace** old cells
- for **asexual** reproduction.

Before cells divide, the doubling up of chromosomes means that DNA has to copy or **replicate** itself.

During mitosis, the chromosomes in the cell are copied to produce two cells that are **genetically identical**.

HT *N.B. For simplicity, only 2 of the 23 pairs of chromosomes in humans are shown.*

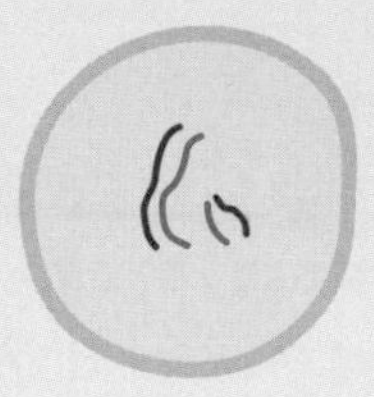

Parental cell with two pairs of chromosomes.

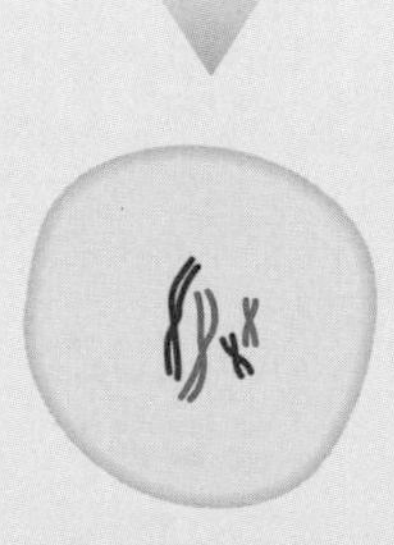

Each chromosome replicates itself, and lines up on the equator of the cell.

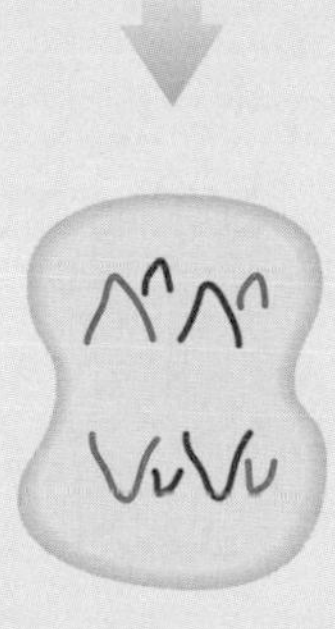

The replicas separate from the originals and move to opposite poles of the cell. The cell then divides for the only time.

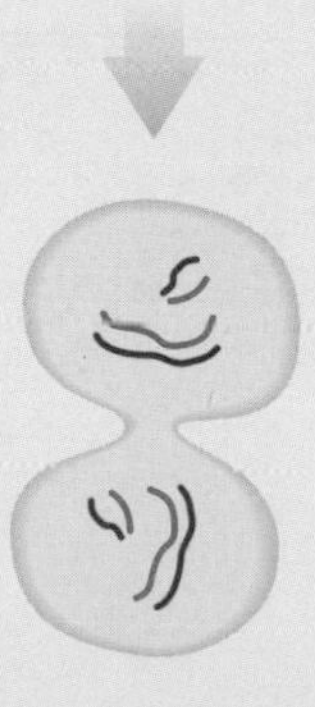

Each 'daughter' cell has the same number of chromosomes as the parental cell and contains the same genes as the parental cell.

## HT Genetic Instructions

A cell can make an exact copy of its DNA molecule in the following way:

1. The double helix 'unzips'.
2. New bases pair up with the exposed bases on each strand.
3. An enzyme bonds the new bases together to form complementary strands.
4. Two identical strands of DNA are formed.

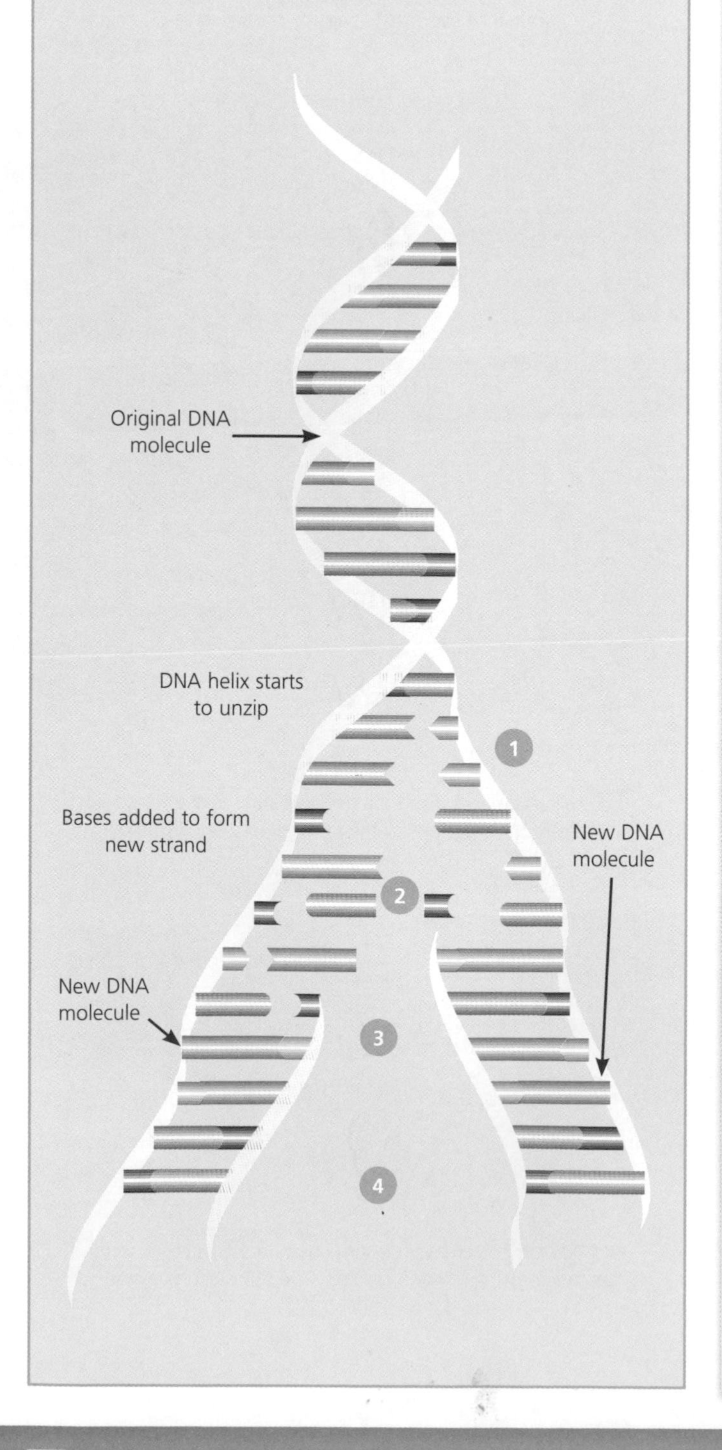


## Meiosis

**Meiosis** is another type of cell division, which occurs in the testes and ovaries. The cells in these organs divide to produce **gametes** (sex cells: eggs and sperm) for sexual reproduction. Gametes are **haploid** cells, which means they have only one copy of each chromosome.

HT During meiosis, a diploid cell divides twice to produce four haploid cells, with genetically different sets of each chromosome.

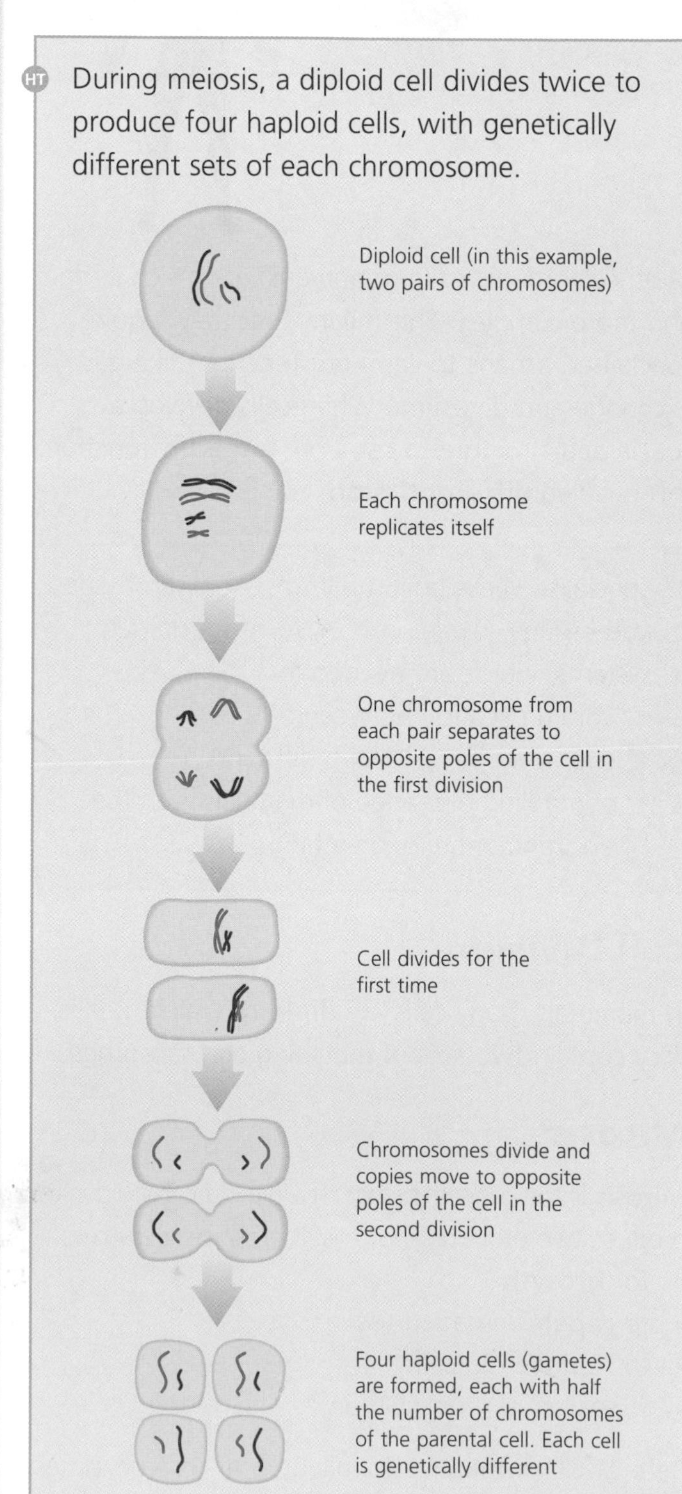

## Fertilisation

**Fertilisation** occurs during sexual reproduction. Two gametes (an egg and a sperm) join or fuse together to form a diploid **zygote** (fertilised egg), with two sets of chromosomes.

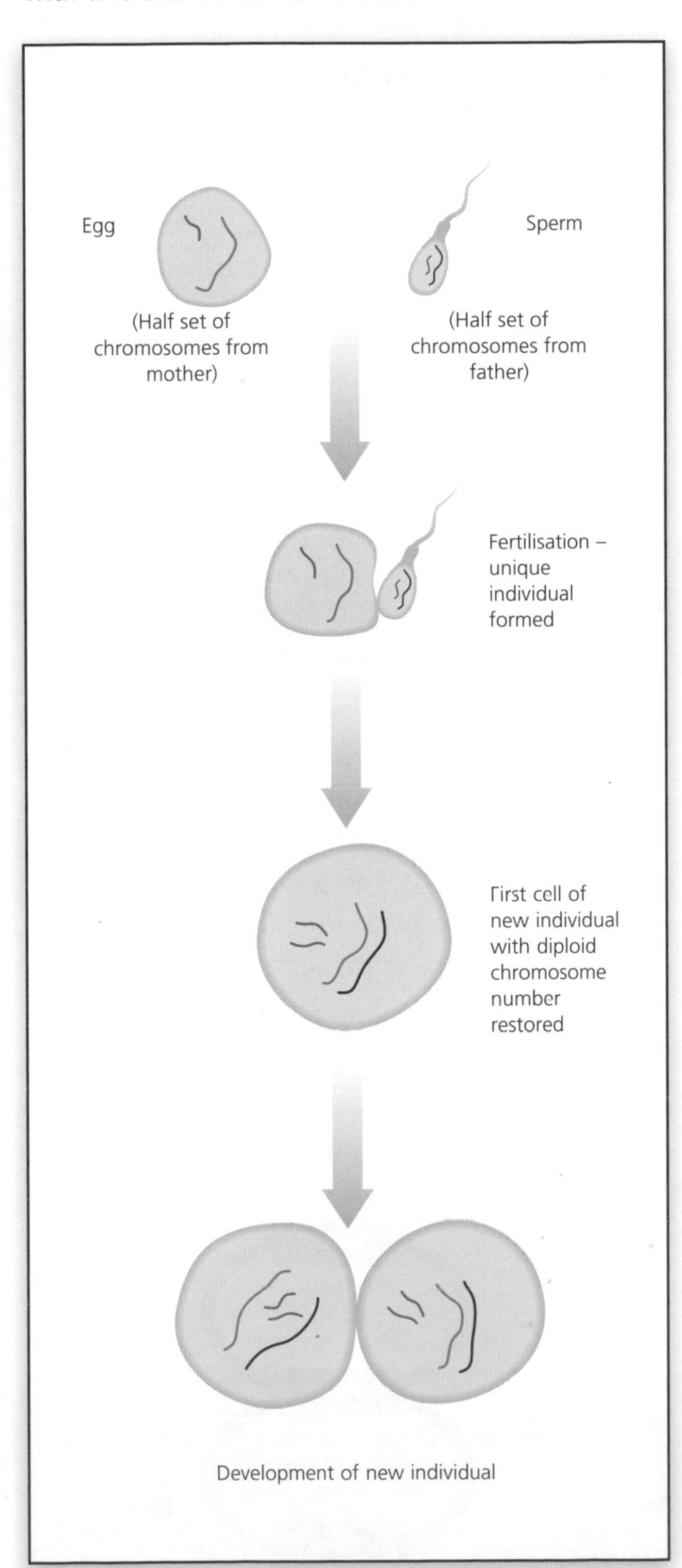


## Gametes

Gametes are specialised haploid cells. This means they have only one copy of each chromosome (half the amount of most body cells). In humans, the gametes are called sperm and egg.

### Special Adaptations of the Sperm Cell

A sperm is a tiny cell with a short life span. It is amazingly mobile because of its tail, and contains many mitochondria in its neck to supply the energy needed for swimming. Its small size is also an adaptation for swimming.

On contact with the egg, its acrosome (a cap-like structure on the sperm's 'head') bursts. This releases enzymes that digest the egg cell's membrane, allowing the sperm nucleus, containing one set of genes from the father, to enter.

Sperm are produced and released in vast numbers to increase the chance of fertilisation occurring.

**Sperm**

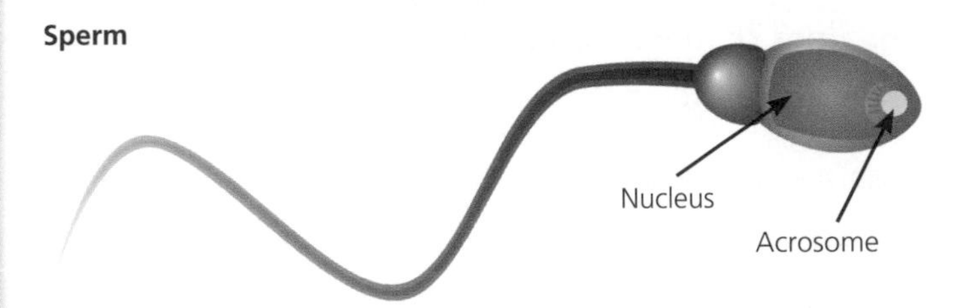


## Variation Through Sexual Reproduction

Sexual reproduction promotes variation because:

- the gametes are produced by meiosis, which 'shuffles' the genes and chromosomes
- gametes fuse randomly, with one of each pair of alleles for a gene coming from each parent
- the alleles in a pair may be the same or different, producing different characteristics.

## The Blood

Blood transports food and oxygen to cells and removes waste products from the cells. It also forms part of the body's defence mechanism. It has four components:

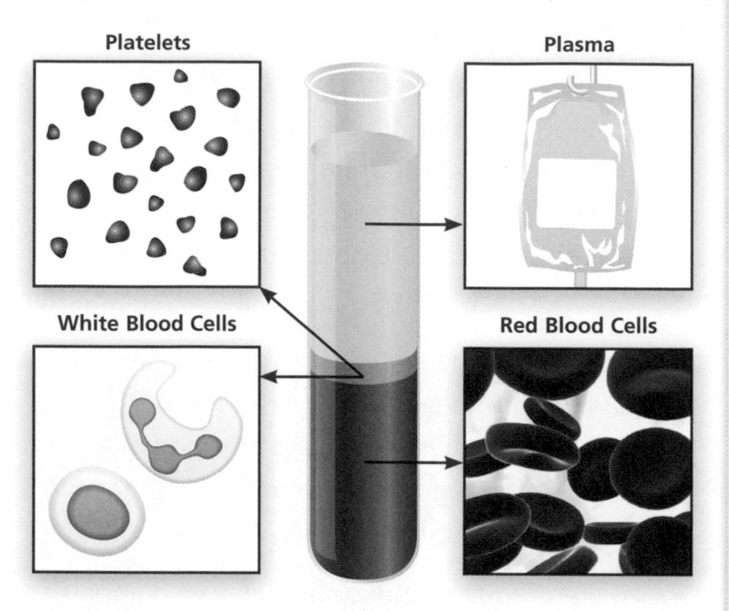


**Platelets** are tiny pieces of cell, which have no nucleus. They clump together when a blood vessel becomes damaged and form a meshwork of fibres to produce a clot.

**Plasma** is a straw-coloured liquid which transports:
- carbon dioxide from the cells to the lungs
- glucose and other digested food products from the small intestine to the cells
- waste products (e.g. urea) from the liver to the kidneys
- hormones to the target organs
- antibodies around the body to fight disease.

**White blood cells** protect the body against disease. Some have a flexible shape, which enables them to engulf invading microorganisms; others produce antibodies to attack the microorganisms.

**Red blood cells** transport oxygen from the lungs to the tissues. They are small and flexible, so they can pass through narrow blood vessels. They have no nucleus, so they can be packed with **haemoglobin** (the red pigment that carries oxygen). The shape of the cells increases their surface area for transferring oxygen.

HT The small size and biconcave shape of red blood cells gives them a large surface area in relation to their volume for transferring oxygen. When the cells reach the lungs, oxygen diffuses from the lungs into the blood. The haemoglobin molecules in the red blood cells bind with the oxygen to form **oxyhaemoglobin**.

**haemoglobin + oxygen ⇌ oxyhaemoglobin**

The blood is then pumped around the body to the tissues, where the reverse reaction takes place. This releases the oxygen so that it can diffuse into the cells.

## The Circulatory System

The **heart** pumps blood around the body in the blood vessels, transporting materials to and from the tissues:
- **Arteries** transport blood away from the heart.
- **Veins** transport blood towards the heart.
- **Capillaries** are involved in exchanging materials with the tissues.

The human **circulatory system** is a **double** circulatory system i.e. it consists of two loops:
- One loop carries blood from the heart to the lungs, and then back to the heart (pulmonary).
- One loop carries blood from the heart to all other parts of the body, and then back to the heart (systemic).

Blood pumped out of the heart into the arteries is under much higher pressure than the blood returning to the heart in the veins. It is this pressure difference that causes blood to flow around the body.

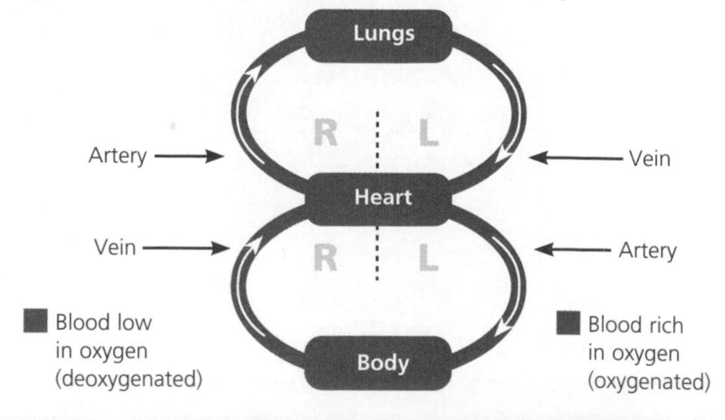

The advantage of a double circulatory system (common to all mammals) is that blood is pumped at a higher pressure because it receives two 'boosts' from the heart for every complete circuit of the body. This results in a greater rate of flow to the body tissues.

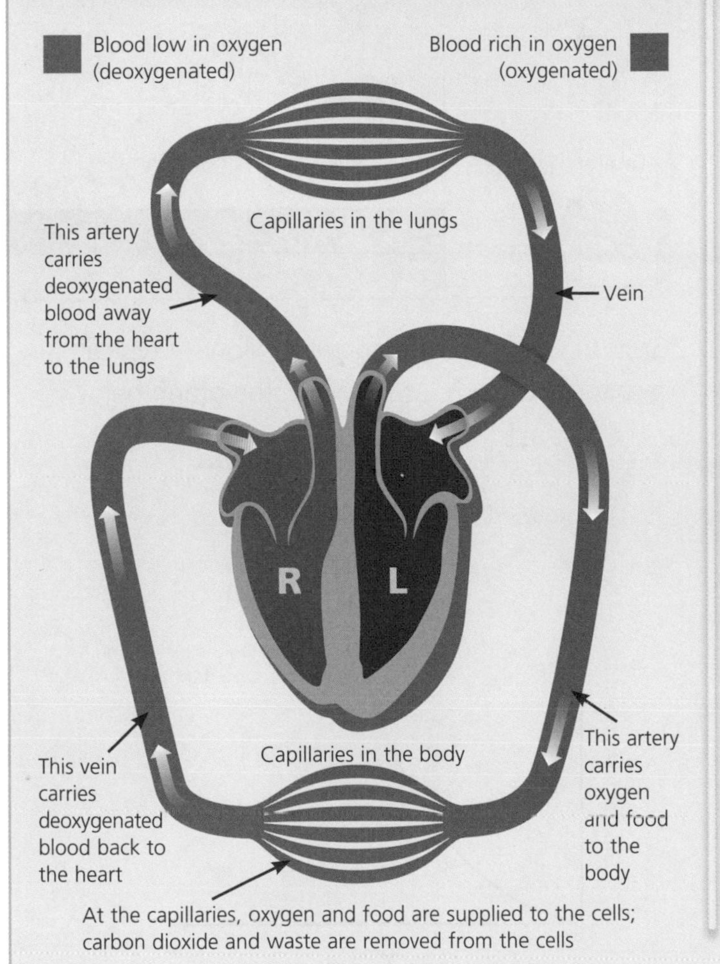


**Arteries** have a thick wall made of elastic fibres and muscle fibres to cope with the high pressure. The **lumen** (space inside) is small compared to the thickness of the walls. There are no valves.

**Veins** have thinner walls made up of a thinner layer of muscle fibres and fewer elastic fibres. The lumen is much bigger compared to the thickness of the walls and there are valves to prevent the backflow of blood.

**Capillaries** are narrow vessels with walls that are just one cell thick. These microscopic vessels connect arteries to veins. They are the only blood vessels that have permeable walls, to allow the exchange of substances between cells and blood.

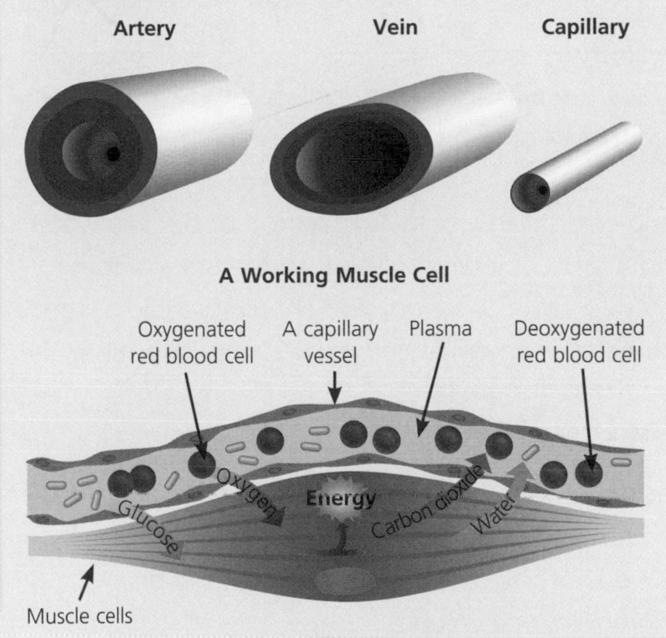


## The Heart

Most of the wall of the heart is made of **muscle**. It has four chambers. The lower chambers, called **ventricles**, are large and muscular because they need to contract to pump blood out of the heart. The right ventricle pumps blood to the lungs. The left ventricle is more muscular because it has to pump blood around the whole body. The upper chambers, called **atria**, are smaller and less muscular. They receive blood coming back to the heart through the veins. Valves in the heart make sure that the blood flows in the correct direction through the heart.

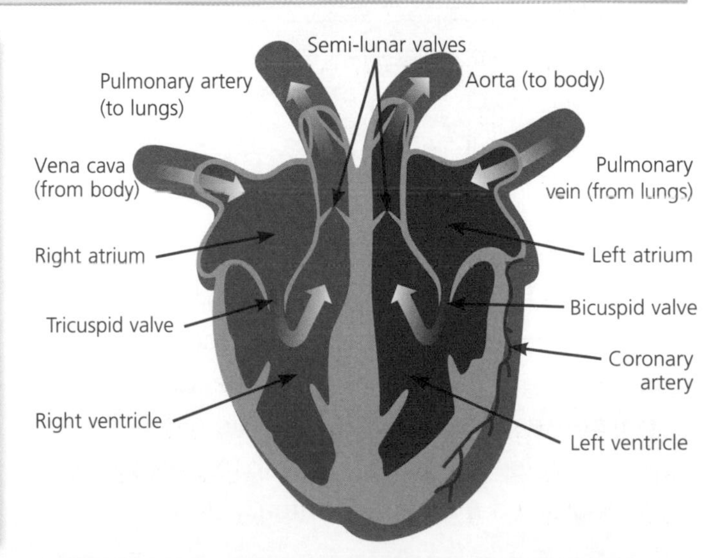

## Plant Cells

Plant and animal cells contain a nucleus, cell membrane, mitochondria and cytoplasm. Plant cells also contain chloroplasts, a cellulose cell wall and a vacuole.

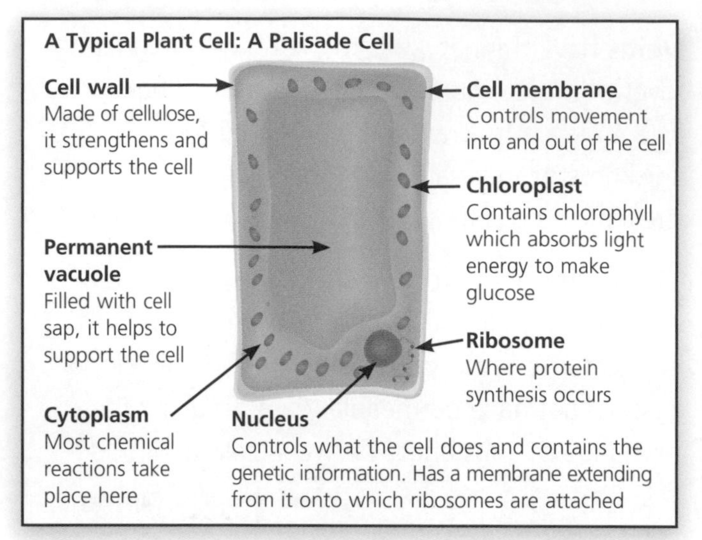


To see the parts of a plant cell, use tweezers to peel a thin layer of skin tissue from an onion. Place the onion tissue onto a microscope slide on top of a drop of distilled water. Add a drop of distilled water and a drop of iodine solution to the tissue and carefully cover the slide. You should be able to see the parts of an onion cell at ×100 magnification.

## Bacterial Cells

Bacteria are very small single-celled organisms. Animal and plant cells are 10 to 100 times bigger. Bacterial cells do not have a 'true' nucleus, mitochondria, chloroplasts or a vacuole.

HT Plant cells keep their DNA inside the nucleus but bacterial cells do not have a nucleus so their DNA floats as circular strands.

## Growth

Growth can be measured as an increase in height or mass. Animals grow by increasing the number of cells in the early part of their lives. The cells specialise or **differentiate** at an early stage into different types of cell to form tissues and organs. All parts of animals grow, but growth eventually stops.

Plants grow by cell division and **cell enlargement**. They gain height mainly through cell enlargement. Many plant cells retain the ability to differentiate so they can grow continually throughout their life. Growth only happens in the roots and the shoot tips – areas called **meristems**.

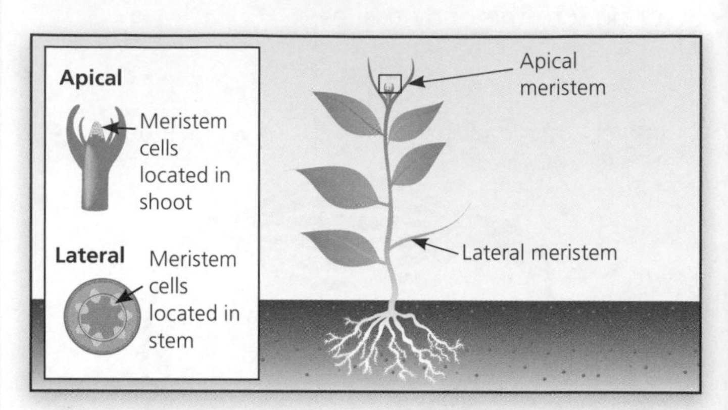


Growth in organisms generally follows a typical 's-shaped' pattern, as shown in the graph below:

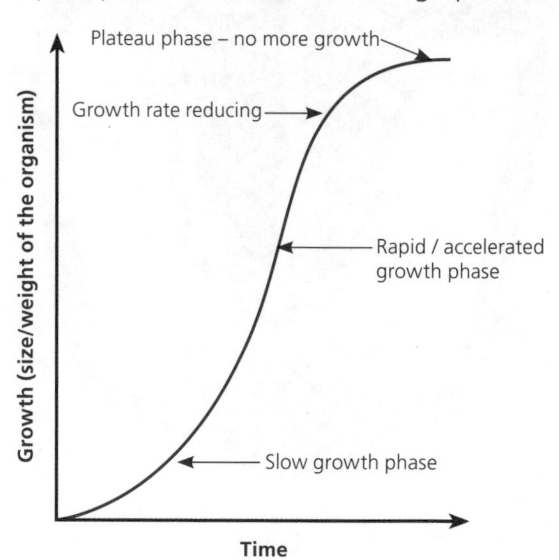


## Measuring Growth

Growth can be measured as an increase in length or as an increase in wet or dry mass. The best measure of growth is dry mass.

HT

| Method | Advantage | Disadvantage |
|---|---|---|
| Length | Easy and rapid measurement | Increase in mass might occur with no increase in length |
| Wet Mass | Not destructive, is relatively easy to measure | Water content of living tissue can be very variable and may give a distorted view overall |
| Dry Mass | Most accurate method | Destructive as removal of water kills organism |

## Human Growth

Humans grow fastest during infancy and early childhood, and have another growth spurt in adolescence (puberty). As adults (18 and over), humans grow very little, if at all. The graph below shows how many centimetres males and females grow on average each year for the first 20 years of their life.

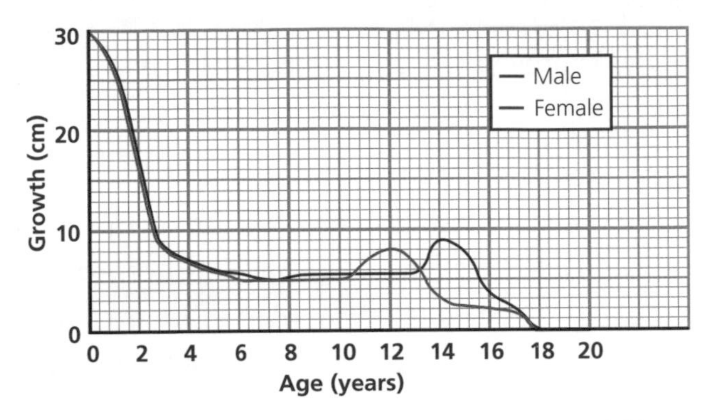


A human foetus in the uterus grows very quickly. Different body parts develop at different rates. The head and brain grow rapidly at first to coordinate the complex growth of the rest of the body.

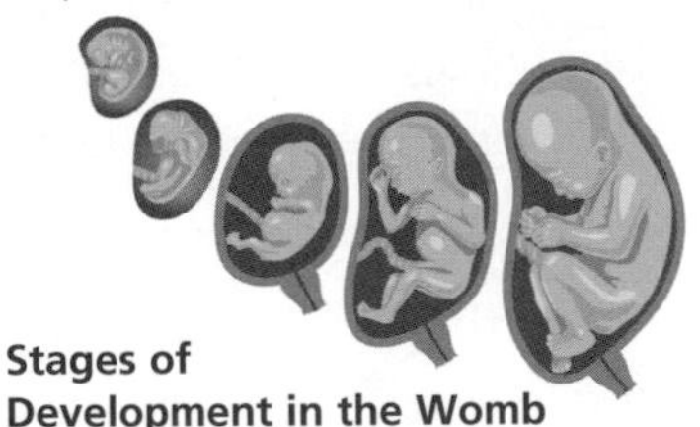

Stages of Development in the Womb

## Differential Growth

During growth there may be times when some parts of the body grow at a different rate to others. For example, a one-week old embryo will have almost a third of its body taken up with heart tissue. This is because the circulation of blood is vital to the growth and development of the embryo and therefore needs to be prioritised.

## Growth in Plants

**Cell division** is mainly restricted to the tips of the roots and shoots. The cells specialise into **xylem**, **phloem** and a variety of other types of cell.

Unlike animal cells, plant cells retain the ability to differentiate or specialise throughout their lives. If you take cuttings from a section of stem, some of the cells will be able to turn into root cells.

## Stem Cells

**Stem cells** are undifferentiated cells, which can specialise and develop into different types of cells, tissues and organs.

Stem cells found in embryos can specialise into any type of cell.

Stem cells found in some adult tissues such as bone marrow can only specialise into a limited variety of cells, e.g. blood or bone cells.

There are a lot of potential uses for stem cells. Scientists believe they could be useful for:

- **research** – to investigate how cell division goes wrong in diseases like cancer
- **drug testing** – to test the safety of new medicines
- **transplants** – to replace damaged or diseased cells, or to grow new organs for transplantation.

In order to carry out research, scientists need to obtain embryos in large numbers and to grow stem cells in the laboratory.

At the moment, the embryos used come from **IVF** (*in vitro* fertilisation) treatments for infertile couples.

Only a few of the embryos produced by IVF treatment are implanted back into the woman, so scientists can use the remaining ones with the couple's consent.

### Ethical Conflicts

There are opposing views about obtaining embryos for stem cell research.

One side of the argument is that the embryos left over from IVF treatment would be destroyed in any case, so stem cell research is a good use for them.

The other viewpoint is that embryos have the potential to become a human being, so it is wrong to experiment on them.

## Selective Breeding

Farmers and dog breeders have used the principles of **selective breeding** for hundreds of years by keeping the best animals and plants for breeding, e.g. the spottiest dogs have been bred through the generations, to eventually get Dalmatians.

The process of selective breeding is as follows:

- Select the desired characteristics in parents.
- Allow the individuals to breed (or cross-pollinate if you are dealing with plants).
- Select desired offspring and allow these to become parents of the next generation.

This process has to be repeated many times to get the desired results.

## Examples of Selective Breeding

### Modern Vegetables

The diagram below illustrates how three of our modern vegetables have come from a single ancestor by selective breeding.

*N.B. It can take many, many generations to get the desired results.*

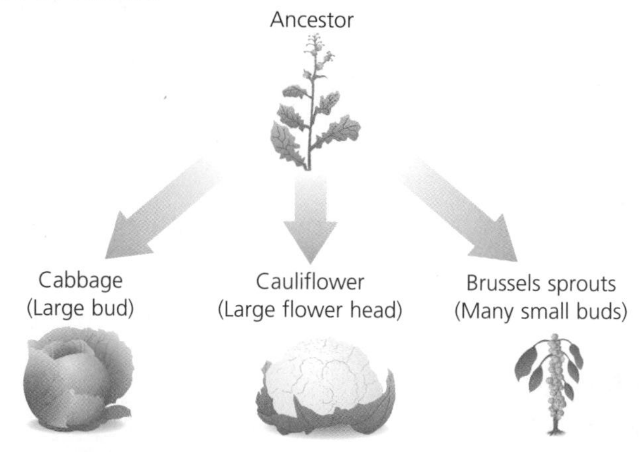


### Modern Cattle

Selective breeding can contribute to improved yields in cattle, for example:

- **Quantity of milk** – years of selecting and breeding cattle that produce larger than average quantities of milk has produced herds of cows that produce high volumes of milk daily.
- **Quality of milk** – as a result of selective breeding, Jersey cows produce milk that is rich and creamy and can therefore be sold at a higher price.
- **Beef production** – the characteristics of the Hereford and Angus varieties have been selected over the past 200 years or more. They include hardiness, early maturity, high numbers of offspring, and the swift, efficient conversion of grass into body mass (meat).

Similarly, improved yields in crops have been obtained by selective breeding.

### Advantages and Disadvantages

Selective breeding results in an organism with the 'right' characteristics for a particular function. In farming and horticulture it is a more efficient and economically viable process than natural selection.

HT However, intensive selective breeding reduces the gene pool – the range of alleles in the population decreases so there is less variation. This reduces the species' ability to respond to environmental change and limits the opportunities for further selective breeding. It can also lead to an accumulation of harmful recessive characteristics (in-breeding).

## Genetic Engineering

Because all living organisms use the same basic genetic code (DNA), genes can be transferred from one organism to another in order to deliberately change the recipient's characteristics. This process is called **genetic engineering** or **genetic modification (GM)**.

Altering the genetic make-up of an organism can be done for many reasons, for example:

- **To improve resistance to herbicides**, e.g. soya plants are genetically modified by inserting a gene that makes them resistant to a herbicide. When the crop fields are sprayed with the herbicide, only the weeds die leaving the soya plants without competition so they can grow better. Resistance to frost or disease can also be genetically engineered.
- **To improve the quality of food**, e.g. the genes responsible for producing beta-carotene

(which is converted into vitamin A in humans) are transferred from carrots to rice plants. People whose diets lack vitamin A can then get beta-carotene from rice.
- **To produce a substance you require**, e.g. the gene for human insulin can be inserted into bacteria to make insulin on a large scale to treat diabetes.

Genetic engineering allows organisms with new features to be produced rapidly. It can also be used to make biochemical processes cheaper and more efficient. However, the transplanted genes may have unexpected harmful effects.

In the future it may be possible to use genetic engineering to change a person's genes and cure certain disorders e.g. cystic fibrosis. This is an area of research called **gene therapy**.

## HT Producing Insulin

The following method is used to produce insulin:

1. The human gene for insulin production is identified and removed using a **restriction enzyme**, which cuts through the DNA strands in precise places.
2. The same restriction enzyme is used to cut open a ring of bacterial DNA (**a plasmid**). Other enzymes are then used to **insert** the section of human DNA into the plasmid.
3. The plasmid is reinserted into a bacterium which starts to divide rapidly. As it divides it **replicates** the plasmid.
4. The bacteria are cultivated on a large scale in **fermenters**. Each bacterium carries the instructions to make insulin. When the bacteria then make the protein, commercial quantities of insulin are produced.

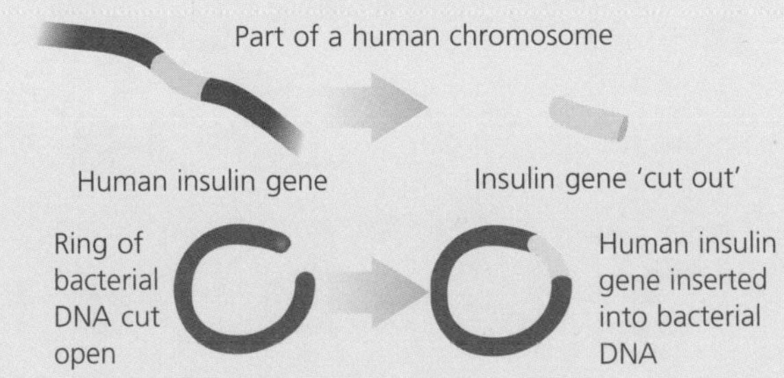


## Ethical Considerations

There are many benefits of genetic engineering, such as producing disease-resistant crops and higher yields, and potentially faulty genes could be replaced to reduce certain diseases. However, there are some concerns – for example, that:
- genetically modified plants may cross-breed with wild plants and release their new genes into the environment
- GM foods may not be safe to eat in the long term
- rather than just replacing 'faulty' genes, parents may want to engineer the genetic make-up of their child (creating 'designer babies').

Currently, genetic screening takes place in certain circumstances, but there is a fear that screening could become more widespread and unborn children could be genetically screened and aborted if their genetic make-up is 'faulty'.

## HT Arguments About Gene Therapy

Gene therapy involves the use of body cells or gametes. With gametes there are ethical issues raised because so-called 'germ-line gene therapy' is in its infancy. Opponents say:
- Very little is known about gene regulation or the mechanisms of embryological development. The results of premature use of such techniques could be worse than the diseases they are meant to cure.
- It may not just be diseases that are removed from the population. Conditions like myopia, racial variations like skin colour and height might be targeted.
- Reduction in genetic diversity of the human gene pool could increase vulnerability to new diseases.

There are also concerns that insurance companies could genetically screen applicants and refuse to insure people at a higher risk of illness, preventing them from being able to drive a car or buy a home.

# B3 Cloning

## Asexual Reproduction in Plants

Plants can reproduce **asexually** (i.e. without the fertilisation of sex cells) and many do so naturally. In asexual reproduction, a cell divides by **mitosis** to produce two identical cells; each new cell continues to divide and develop to produce genetically identical individuals or **clones**.

The strawberry plant, potato plant and spider plant can all reproduce in this way.

Potato plants produce **tubers** (i.e. the potatoes themselves). The tubers produce several buds, from which new plants grow.

Strawberry plants produce **runners**, from which new plants root and grow.

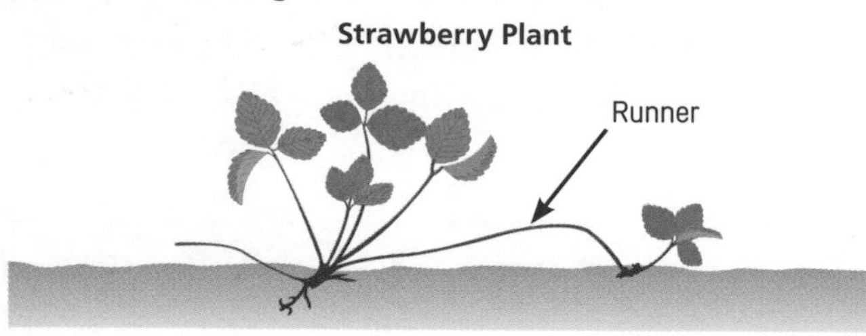


Spider plants produce runners (stolons) from which new individual plants grow.

**Spider Plant Runners (Stolons)**

Runner – a rooting side branch

New individual established

New individual now independent

## Taking Cuttings

When a gardener has a plant with all the desired characteristics, he/she may choose to produce lots of them by taking stem, leaf or root cuttings. The cuttings are grown in a damp atmosphere until roots develop.

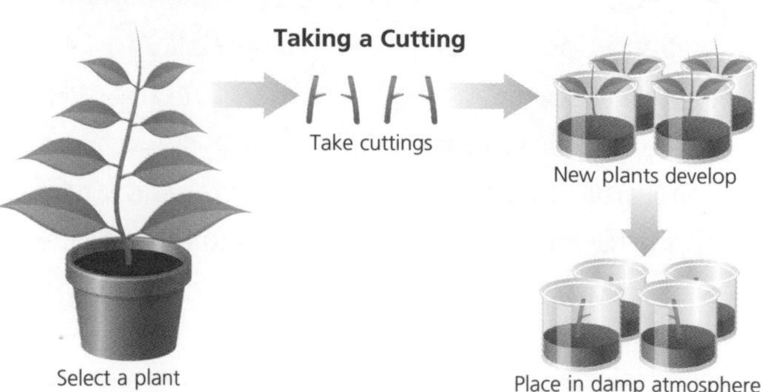


## Commercial Cloning of Plants

In modern horticulture, prize plant specimens are cloned to produce thousands more, which can be sold. The advantages and disadvantages of this process are listed in the following table:

| Advantages |
| --- |
| • You can be sure of the characteristics of the offspring because they will all be genetically identical to the parent.<br>• It is possible to mass-produce large numbers of plants, which may have been more difficult, or taken more time, to grow from seed. |
| **Disadvantages** |
| • If the plants become susceptible to disease or sensitive to a change in the environment, all of the plants will be affected.<br>• The reduction in genetic variation reduces the potential for further selective breeding. |

## HT Cloning by Tissue Culture

To produce offspring that are genetically identical to the parent plant and to each other, horticulturalists follow this method:

1. Select a parent plant with the characteristics that you want.
2. Scrape off a large number of small pieces of tissue into several beakers containing nutrients and hormones.
3. A week or two later there will be lots and lots of genetically identical plantlets growing. The same process can then be repeated.

HT This whole process must be **aseptic** (i.e. carried out in the absence of harmful bacteria), otherwise the new plants will rot.

Remember, many older plant cells are still able to differentiate or specialise, whereas animal cells lose this ability as they grow older. This means that cloning plants is easier than cloning animals.

## Cloning Animals

Identical twins are naturally occurring clones in animals. A single fertilised egg forms an embryo that splits into two at an early stage. The two individuals develop from the same fertilised egg and have identical genetic make-up.

Artificial cloning of animals is now used quite widely. The most famous animal clone is Dolly the sheep – the first mammal to be successfully cloned from an adult body cell. Dolly was produced using the process of **nuclear transfer**. The technique involved taking a nucleus from a sheep's body cell and then placing it into an egg cell that had its nucleus removed.

A cloning technique called **embryo transplantation** is now commonly used in cattle breeding:

1. Sperm is collected from a bull with desirable characteristics.
2. A selected cow is artificially inseminated with the bull's sperm.
3. The fertilised egg develops into an embryo which is removed from the cow at an early stage.
4. In the laboratory, the embryo is split to form several clones.
5. Each clone is transplanted into a cow who will be the surrogate mother to the new calf.

## Animal Organ Donors

There is a shortage of human organ donors for transplants. One possible solution is to use animal organs.

Animal organs would normally be rejected and destroyed by the human immune system. However, animal embryos could potentially be genetically modified so that they will not be rejected. They could then be cloned to produce a ready supply of identical organ donors.

HT Animal organ donors could solve the problem of waiting lists for human transplants. But there are concerns that infections might be passed from animals to humans, and there are ethical issues concerning animal welfare and rights.

## Uses of Cloning

There are several uses for cloning:

- It is possible to clone human embryos in the same way that animals are cloned. This technique could be used to provide stem cells for medical purposes.
- The mass production of animals with desirable characteristics.

## Human Cloning

It is possible to clone human embryos in the same way that animals are cloned.

This technique could be used to provide stem cells for medical purposes. However, if the embryos were allowed to develop they would produce human clones, and this type of research is currently illegal beyond 14 days.

There are major concerns about cloning humans:

- The cloning process is very unreliable – the vast majority of cloned embryos do not survive.
- Cloned animals seem to have a limited life span and die early.
- The effect of cloning on a human's mental and emotional development is not known.
- Religious views say that human cloning is wrong.
- Using human embryos and tampering with them is controversial.

## Adult Cell Cloning

The following method was used to produce a cloned sheep (Dolly):

1. A nucleus was taken from an udder cell of an adult sheep, and the nucleus was removed from an egg cell of a female sheep.
2. The nucleus from the udder cell was then inserted into the empty egg cell using a brief pulse of electric current to make the embryo divide.
3. The resulting embryo was placed into the uterus of a surrogate mother sheep.
4. The embryo developed into a foetus and was born as normal. The offspring produced (Dolly) was a clone of the sheep that the nucleus came from.

**Cloning**

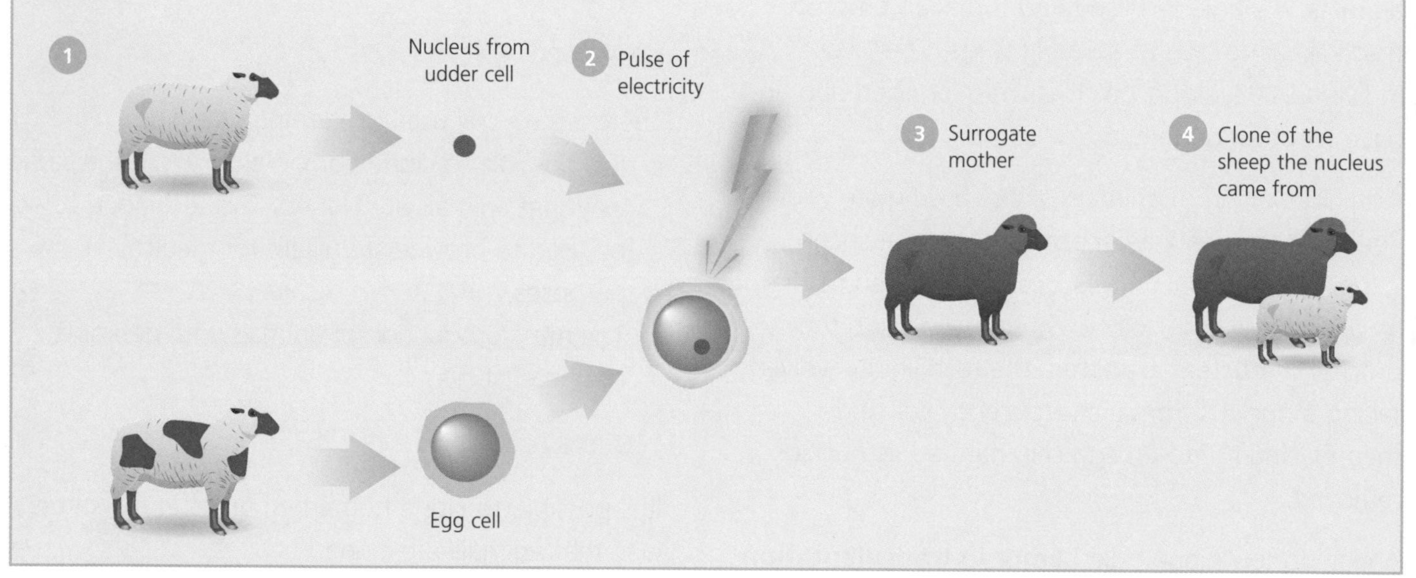


## Benefits and Risks of Cloning

There are benefits and risks associated with cloning technology:

| Benefits | Risks |
|---|---|
| • Genetically identical cloned animals will all have the same characteristics.<br>• It is possible for a farmer to choose the sex and timing of birth.<br>• Top-quality bulls and cows can be kept for egg and sperm donation, while other animals can be used to carry and give birth to the young. | • Just like selective breeding, cloning reduces genetic variation in the herd.<br>• There is the potential for accumulating inherited diseases.<br>• There are some animal welfare concerns that cloned animals may not be as healthy or live as long as 'normal' animals. |

1. The table below charts Thomas's growth in mass from birth to age 2.

| Age (months) | 0 | 3 | 6 | 9 | 12 | 15 | 18 | 21 | 24 |
|---|---|---|---|---|---|---|---|---|---|
| Mass (kg) | 2.2 | 3.4 | 4.2 | 5.0 | 5.6 | 6.0 | | 7.2 | 7.4 |

The graph shows the growth rate of another baby, Joshua. Joshua had no health problems from birth to age 2.

**a)** Plot a graph of Thomas's growth data and create a growth curve. **[2]**

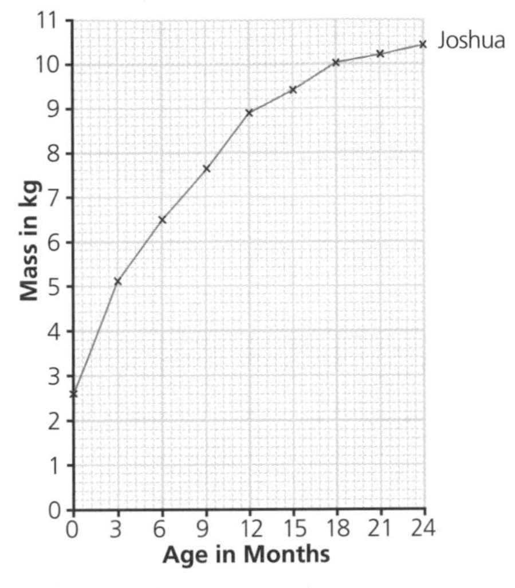


**b)** One value in the table has been left blank. Use your graph to predict this value. **[1]**

**c)** Between which ages did Thomas grow fastest? **[1]**

**d)** Thomas's growth curve is very different to Joshua's. He was undernourished as a baby due to a problem with his digestive system. At 3 years of age he recovered from his ailment. Describe what would happen to his growth curve between the ages of 3 and 7 years. **[1]**

**e)** Thomas's parents planted an oak tree in their garden when he was born. Describe how the increase in mass would be different for Thomas and the oak tree between the ages of 12 and 50. **[2]**

2. This question is about the heart.

**a)** On the diagram, label the ventricle with an 'X'. **[1]**

HT **b)** A science student observes some sections of arteries and veins under the microscope. Describe **two** differences she might see between the vessels. **[2]**

HT 3. This diagram represents the amino acid sequence in a short length of protein.

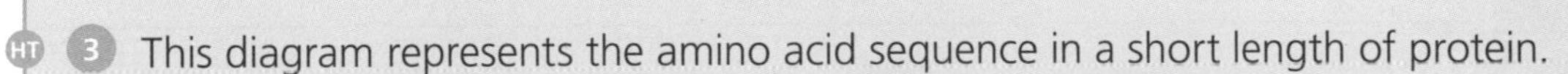

**a)** Which group of chemicals in DNA code for the sequence? **[1]**

**b)** The protein is a digestive enzyme which works in the mouth. It works well at a pH of 7.8. Explain why the enzyme stops working when it passes into the stomach. **[2]**

# B4 Ecology in the Local Environment

## B4: It's a Green World

This module looks at:

- Identifying organisms using keys and how plants and animals can be sampled in the field.
- Photosynthesis and the uses of its products.
- How leaves are adapted to photosynthesis.
- Diffusion and osmosis and how cells interact with water in plants and animals.
- How plant tissues carry out their transport functions and how transpiration is controlled.
- The ways in which plants obtain the minerals they need.
- Decomposition of materials by microorganisms, and knowledge of decay in preserving food.
- Organic and intensive farming, and hydroponics.

## Sampling Methods

A **population** is the total number of individuals of the same species that live in a certain area, e.g. the number of field mice in a meadow.

The size and distribution of a population can be measured using one or more of the techniques shown below.

HT When sampling, you must make sure you:

- take a big enough sample to make the results a good and reliable estimate – the larger the sample then the more accurate the results
- sample randomly – the more random the sample the more likely it is to be representative of the population.

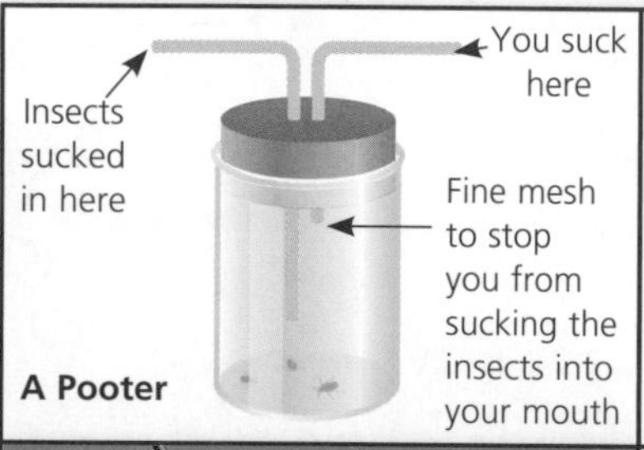


**Pooters**

This is a simple technique in which insects are gathered up easily without harm. With this method, you get to find out which species are actually present, although you have to be systematic about your sampling in order to get representative results and it is difficult to get ideas of numbers.

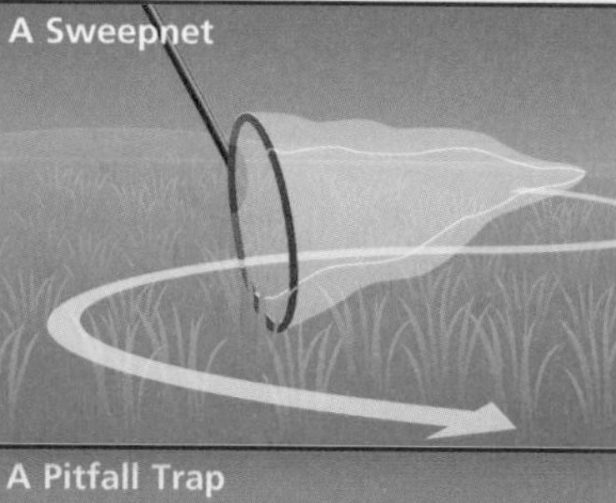


**Sweepnets**

Sweepnets are used in long grass or moderately dense woodland where there are lots of shrubs. Again, it is difficult to get truly representative samples, particularly in terms of the relative numbers of organisms.

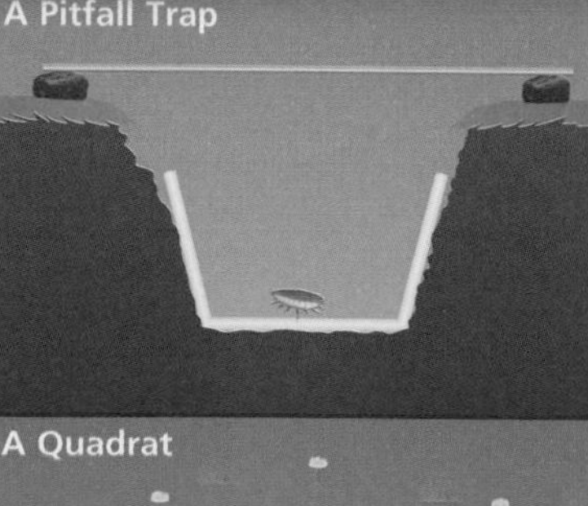


**Pitfall Traps**

Pitfall traps are set into the ground and used to catch small insects, e.g. beetles. Sometimes a mixture of ethanol or detergent and water is placed in the bottom of the trap to kill the samples, and prevent them from escaping. This method can give an indication of the relative numbers of organisms in a given area if enough are used to give a representative sample.

A Quadrat
0.5m
Area = $0.25m^2$
0.5m

**Quadrats**

Quadrats are square frames that typically have sides of length 0.5m. They provide excellent results as long as they are placed randomly. The population of a certain species can then be estimated. For example, if an average of 4 dandelion plants are found in each $0.25m^2$ quadrat, a scientist would estimate that 16 dandelion plants would be found in each $1m^2$, and 16 000 dandelion plants in a $1000m^2$ field.

# Ecology in the Local Environment

## Estimating Animal Populations

Animal populations are difficult to sample and measure because most species are highly mobile and are not easily visible because they are usually hiding from predators. One method is the capture/recapture method, sometimes called the **Lincoln index**:

1. Animals are caught humanely – for example, woodlice are caught in traps overnight. Their number is counted and recorded.
2. The animals are marked in some way – for example, water boatmen (a type of insect) can be marked with a drop of waterproof paint on their upper surface.
3. The marked animals are then released back into the population for a suitable amount of time.
4. A second sample is obtained which will contain some marked animals and some unmarked. The numbers in each group are again counted.

The following formula can then be used to estimate the total population size in the habitat:

$$\text{Population size} = \frac{\text{No. in 1}^{\text{st}}\text{ sample (all marked)} \times \text{No. in 2}^{\text{nd}}\text{ sample (marked and unmarked)}}{\text{No. in 2}^{\text{nd}}\text{ sample which were previously marked}}$$

**Example**

56 mice are caught in woodland and a small section of fur removed from their tail with clippers. The mice are released. The next evening, a further sample of 62 mice are caught. 25 of these mice have shaved tails. What is the total population size?

$$\text{Population size} = \frac{56 \times 62}{25} = \textbf{139 mice}$$

HT Certain assumptions are made when using capture/recapture data. These include:

- No death, immigration or emigration.
- Each sample is collected in exactly the same way without bias.
- The marks given to the animals do not affect their survival rate, e.g. using paint on invertebrates requires care because if too much is added it can enter their respiratory passages and kill them.

## Using Transects

Sometimes an environmental scientist may want to look at how species change across a habitat, or the boundary between two different habitats – for example, the plants found on a river bank as it slopes toward the river. This needs a different approach that is more systematic than random:

- A line such as a tape measure is laid down with regular intervals marked on it.
- A small quadrat is laid next to the line and the number of plants of the different species is estimated or counted. This can sometimes be done by estimating the percentage cover.
- The quadrat is moved along at regular intervals and the plant populations estimated and recorded at each point until the end of the line.

The results can be presented in the form of a kite diagram:

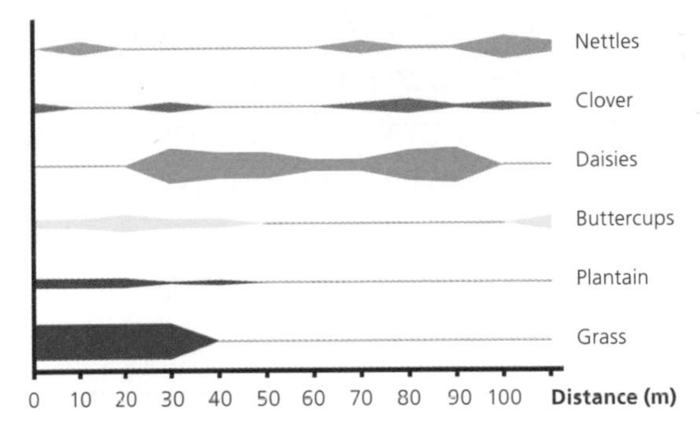


The height of a particular kite shows how abundant that particular species is at that point. By looking at the diagram as a whole it is possible to see how plant species change across the habitat and theories can then be put forward to explain this change.

### HT Zonation

**Zonation** is described as a gradual change in the distribution of a species across a habitat. Changes usually result from an abiotic (physical) factor changing. For example, molluscs such as mussels or periwinkles change in number as you move from the low tide mark up to the high tide mark. Above high tide, they cannot be found at all.

# B4 Ecology in the Local Environment

## Keys

Correctly identifying species in a sample can be difficult. Using keys like this one can help to identify organisms:

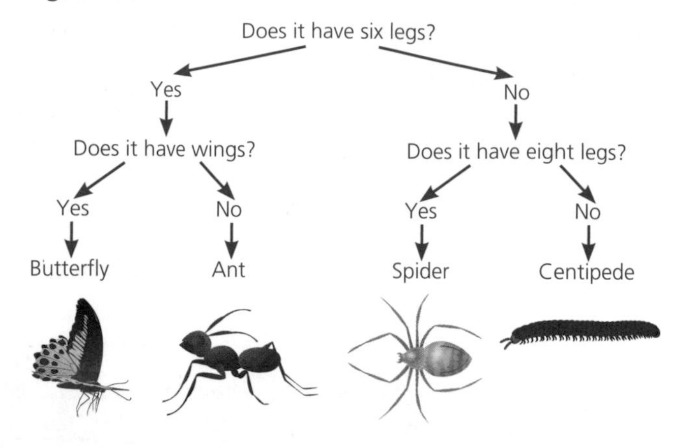


## Some Important Ecological Terms

You will need to know the following terms:

- The **habitat** of an animal or plant is the part of the physical environment where it lives. There are many types of habitat, each with particular characteristics, e.g. pond, hedgerow, coral reef. An organism must be well-suited to its habitat to be able to compete with other species for limited resources. Organisms that are specialised in this way are restricted to that type of habitat because their adaptations are unsuitable elsewhere. Organisms are not usually distributed at random in a habitat. They may often be clumped into specific areas, for example, many freshwater invertebrates are found on the aquatic plants rather than floating in the water. Thus, the distribution of organisms within a habitat is affected by the presence of other living organisms as well as physical factors.
- A **population** is the number of individuals of a species in a defined area.
- A **community** is the total number of individuals of all the different populations of organisms that live together in a habitat at any one time.
- **Biodiversity** is the variety of different species living in a habitat.
- An **ecosystem** is a physical environment with a particular set of conditions, plus all the organisms that live in it. The organisms interact through competition and predation. They are affected by the physical factors present in the ecosystem, e.g. soil type, light intensity. Some ecosystems, such as ocean depths, are still unexplored, with possible undiscovered species.

HT An ecosystem can support itself without any influx of other factors or materials. Its energy source (usually the Sun) is the only external factor.

## Natural and Artificial Ecosystems

**Natural ecosystems** are not man-made (although they may be managed and preserved to provide good habitats for communities of plants and animals). They change over time. Examples of natural ecosystems are native woodlands and lakes.

**Artificial ecosystems** are man-made and carefully controlled to keep the conditions constant. Examples include fish farms and forestry plantations.

Biodiversity in natural ecosystems is always greater than in artificial ones.

HT Natural ecosystems have high **biodiversity** – many different species of plants and animals co-exist in the same environment. Artificial ecosystems are designed for a particular purpose. A market gardener growing a crop in a greenhouse uses **fertilisers**, **weedkillers** and **pesticides** to prevent other organisms from growing alongside his crop. The biodiversity in this ecosystem will be low.

## Comparing Ecological Habitats

Two habitats can be compared by sampling using a quadrat. First the plant and animal species in a $1m^2$ quadrat must be identified and then the population of each species must be counted.

This process is repeated several times to get a large, reliable, random sample.

## Early Ideas about Photosynthesis

How plants gain in mass has been the subject of study for thousands of years. The early Greeks thought plants gained mass by taking in minerals from the soil. Much later, in the 17th century, a Flemish scientist called **Johann Baptista van Helmont** carried out a simple experiment. He dried out 200lb of earth in a furnace (then moistened it with rainwater) and put it in an earthen vessel. He then placed the stem of a willow tree weighing 5lb in the vessel. After five years, the tree had grown to over 169lb. He again dried the earth in the vessel and found it still weighed 200lb.

Van Helmont had noticed that the amount of soil in the vessel he used was equal to the amount of soil at the start of the experiment. Therefore, he deduced that the tree's weight gain had come from water. As the tree had only received water, he proposed that the weight of wood and other plant material had come from water alone. But he missed one crucial factor in his experiment – the effect of gases in the air. We now know that a large proportion of the mass of the tree comes from the carbon dioxide in the air, which, together with water, is turned into carbohydrates through **photosynthesis**.

Further experimental evidence of photosynthesis comes from work done by **Joseph Priestley** in 1771. He put a shoot of mint into a closed glass vessel with a lit candle. The candle extinguished very quickly. After 27 days he relit the candle using sunlight focused by a concave mirror. The candle burned again in the air that previously would not support combustion. We now know that the mint shoot had produced oxygen gas.

## Making Food Using Energy from the Sun

Green plants do not absorb food from the soil. They make their own, using sunlight. This is called **photosynthesis** ('making through light') and it occurs in the cells. The diagram shows what is needed and what is produced during photosynthesis.

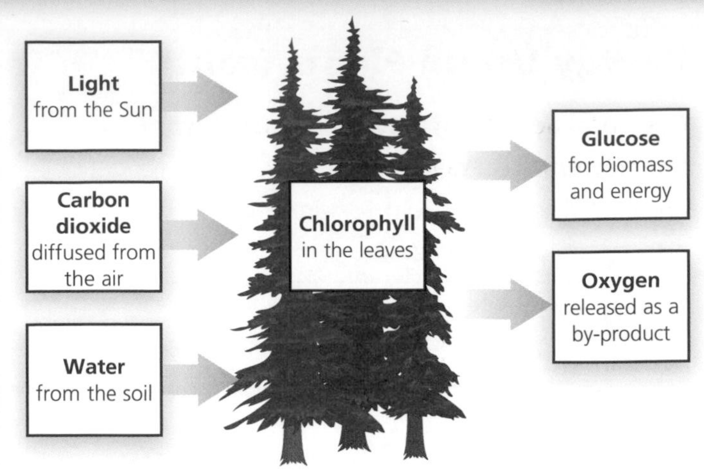


The word and symbol equations for photosynthesis are given below:

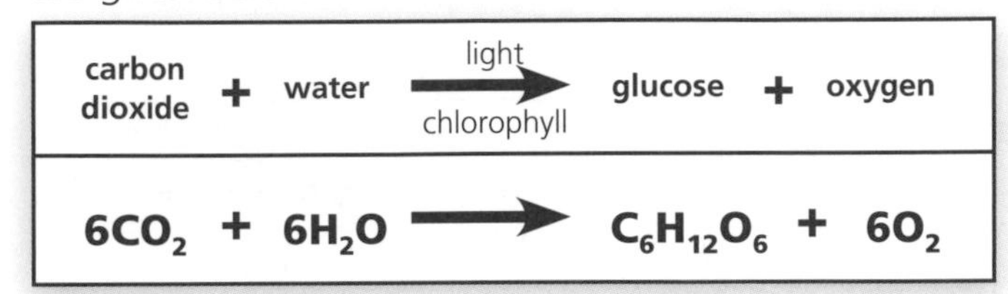


## Energy Use in Plants

The glucose produced in photosynthesis can be used immediately to produce energy through respiration. Some of this energy is used to build up smaller molecules into larger molecules.

### Converting Glucose into Starch

The plant converts **glucose** into **starch** because starch is an insoluble carbohydrate, which can be stored in cells.

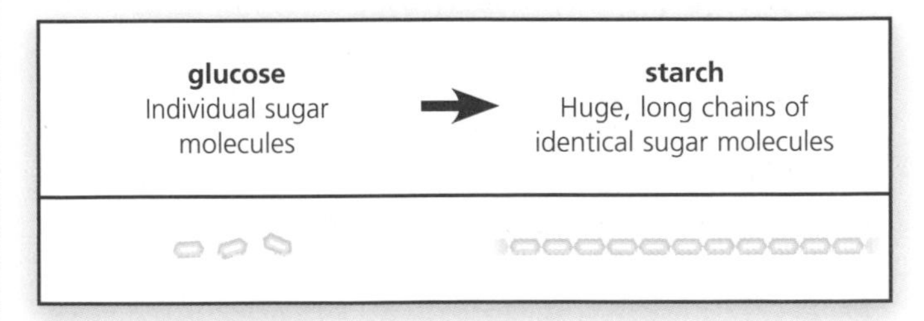


### Converting Glucose into Cellulose

**Cellulose** is needed by the plant for cell walls. It is very similar to the structure of starch, but the long chains are cross-linked to form a meshwork.

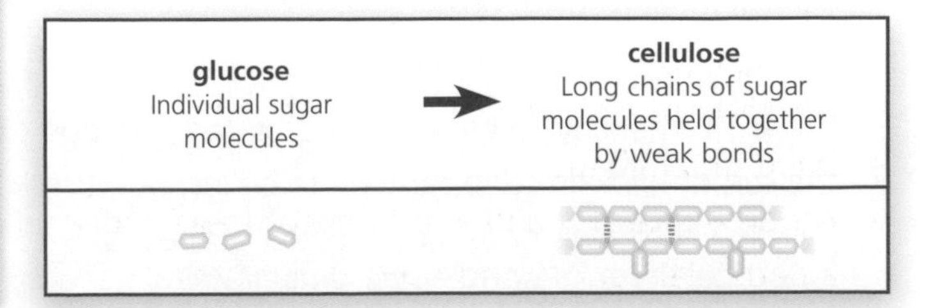

## Energy Use in Plants (cont)

### Converting Glucose, Nitrates and Other Nutrients into Proteins

The plant needs protein for growth and repair and also to make enzymes.

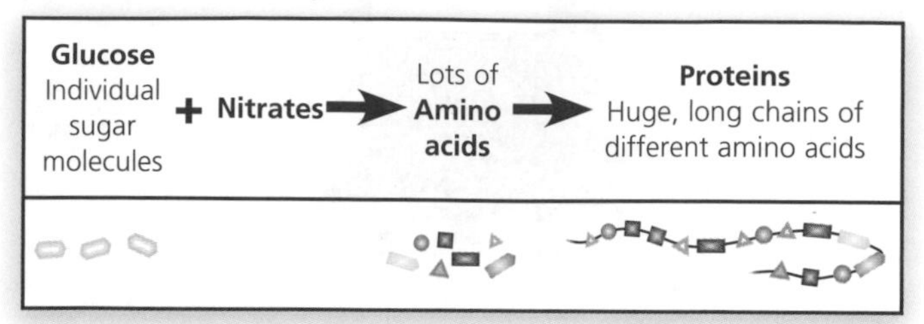


### Converting Glucose into Lipids (Fats or Oils) to Store in Seeds

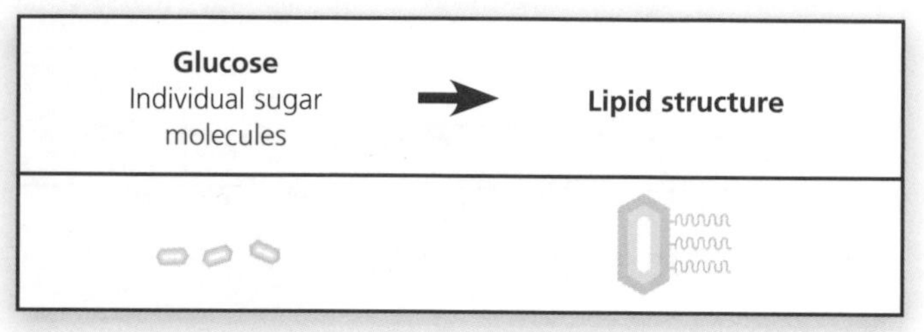


The glucose made in photosynthesis is *transported* as soluble sugars but is *stored* as insoluble starch.

HT Starch is a very useful storage molecule. It is **insoluble** so it does not affect the concentration inside the cells where it is stored. If the cells stored soluble glucose, then the inside of the cells would become very concentrated and water would constantly move in, which would make the cell swell. In addition, the carbohydrate would move away in solution from storage areas.

## Increasing Photosynthesis

Plants need light and warmth to grow. During winter, plant growth slows down, or even stops, due to the lack of sunlight and warmth. Plant growth, or the rate at which the plant photosynthesises, can be artificially increased by growing plants in greenhouses. This enables:

- the temperature to be increased by using heaters
- the light intensity to be increased by using lamps
- the carbon dioxide concentration to be increased by using chemicals or as a by-product of the fossil fuel heaters.

## HT The Chemistry of Photosynthesis

Photosynthesis is a two-stage process:

1. Light energy is used to split water, releasing oxygen gas and hydrogen ions:
   **water → oxygen + hydrogen ions**
2. Carbon dioxide gas then reacts with the hydrogen ions to make glucose and water:
   **carbon dioxide + hydrogen ions → glucose (+ water)**

Early experiments with radioactive isotopes were carried out to discover the source of oxygen in photosynthesis. A plant was given a special type of water called 'heavy water'. It was allowed to absorb this through its roots. Heavy water contains the heavier oxygen atom with an atomic mass of 18 (rather than 16). The plant was allowed to absorb atmospheric carbon dioxide as usual.

When this heavier oxygen atom was traced it was found to be present in the oxygen that was released from the leaves of the plant. Furthermore, it was concluded that none of the evolved oxygen came from the carbon dioxide initially taken in by the plant.

## Respiration in Plants

All living organisms (including plants) **respire** to supply their cells with the energy they need to work, grow and reproduce. Plants take in oxygen and use it to break down glucose to release energy:

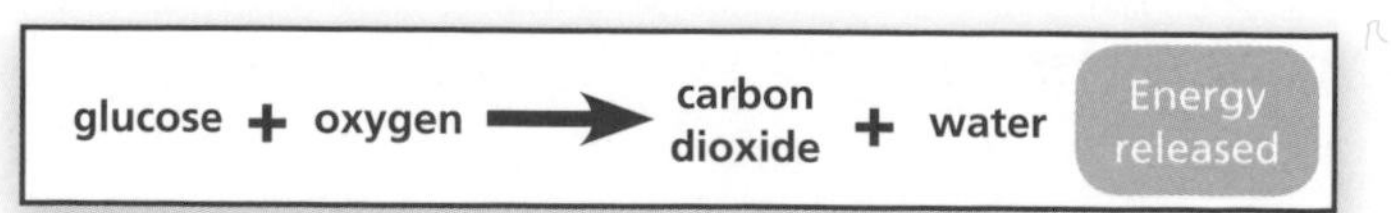


Plants carry out respiration continuously.

## Rate of Photosynthesis

**Temperature**, **carbon dioxide concentration** and **light intensity** can interact to limit the **rate of photosynthesis**. At a particular time, any one of them may be the **limiting factor**.

HT

### Effect of Temperature

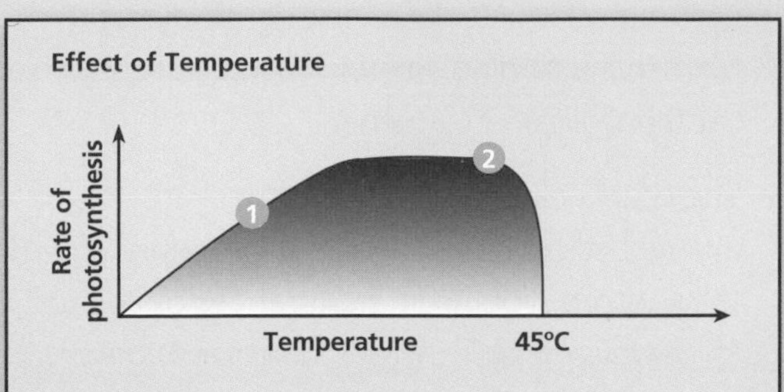


1. As the temperature rises so does the rate of photosynthesis. This means temperature is limiting the rate of photosynthesis.
2. As the temperature approaches 45°C, the enzymes controlling photosynthesis start to be destroyed and the rate of photosynthesis declines to zero.

### Effect of Carbon Dioxide Concentration

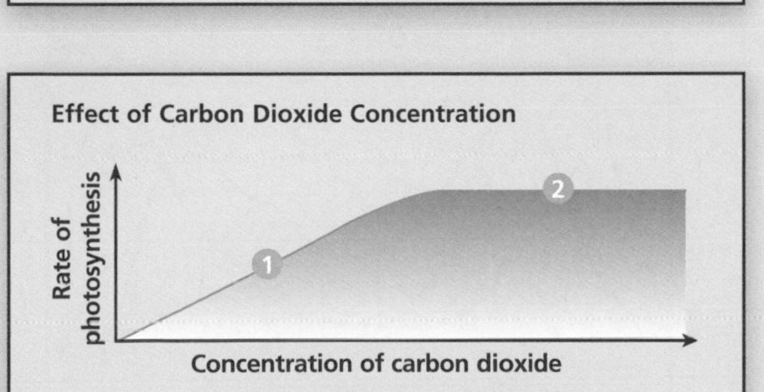


1. As the rate of carbon dioxide concentration rises so does the rate of photosynthesis. So carbon dioxide is limiting the rate of photosynthesis.
2. The rise in carbon dioxide levels now has no effect. Carbon dioxide is no longer the limiting factor; therefore, light or temperature must be.

### Effect of Light Intensity

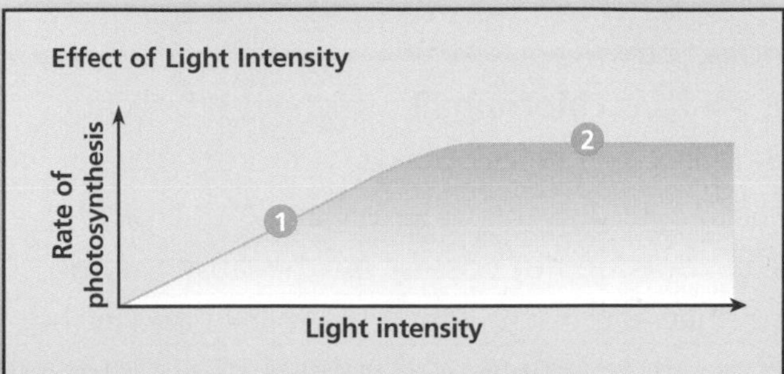


1. As the light intensity increases so does the rate of photosynthesis. This means light intensity is limiting the rate of photosynthesis.
2. The rise in light intensity now has no effect. Light intensity is no longer the limiting factor; carbon dioxide or temperature must be.

## Respiration

During the day, light is available from the Sun so plants are able to photosynthesise; taking in carbon dioxide to make glucose and releasing oxygen as a by-product.

During the night, plants continue to respire, absorbing oxygen and giving out carbon dioxide. Respiration is the reverse of photosynthesis:

- During the day plants respire constantly, but they also photosynthesise at a much greater rate, using up the carbon dioxide produced from respiration and giving out oxygen.
- During the night, the absence of light means that plants cannot photosynthesise, but they continue to respire.

# B4 Leaves and Photosynthesis

## Leaves

Plant cells contain chloroplasts, a large central vacuole and a cell wall. These features make them different from animal cells. The leaves are the food 'factories' of the plant; this is where nearly all **photosynthesis** occurs. They are specially adapted to be super-efficient. Some of the leaf's adaptations are listed below:

- It contains **chlorophyll** (which absorbs light) in millions of **chloroplasts**. (Chloroplasts are rarely found in the other organs of the plant, e.g. roots do not possess any because they do not have access to light.) The chlorophyll in the chloroplasts is able to absorb light from different parts of the spectrum.
- It is broad and flat to provide a **huge surface area** to absorb sunlight for photosynthesis.
- It has a **network of vascular bundles** for support and to transport water to the cells from the root hair cells and remove the products of photosynthesis, i.e. glucose.
- It has a **thin structure**, so the gases (carbon dioxide and oxygen) only have a short distance to travel to and from the cells.
- It has **stomata** (tiny pores) on the underside of the leaf to allow carbon dioxide and oxygen to diffuse in and out for photosynthesis and respiration. The diameter of these stomata are controlled by guard cells.

The carbon dioxide needed for photosynthesis **diffuses** in through the stomata. The oxygen produced by photosynthesis diffuses out. This is called **gaseous exchange**.

## Experiments with Photosynthetic Pigments

Leaves contain chlorophyll and other pigments which absorb different wavelengths (and therefore colours) of light.

HT Chlorophyll is actually a mixture of several pigments including chlorophyll a, chlorophyll b, xanthophylls and carotene.

The chart below shows that when lights of different colours are shone on chlorophyll a and b they absorb slightly different ranges of colours (or wavelengths). However, both tend to absorb colours in the red and violet ends of the spectrum.

When light of different colours is shone on a plant and its rate of photosynthesis is measured, it is found that maximum rates are obtained in the red and violet ends too. These rates are shown in the 'action spectrum' on the graph.

Clearly, the greener colours are reflected, which explains why plants are generally green in colour.

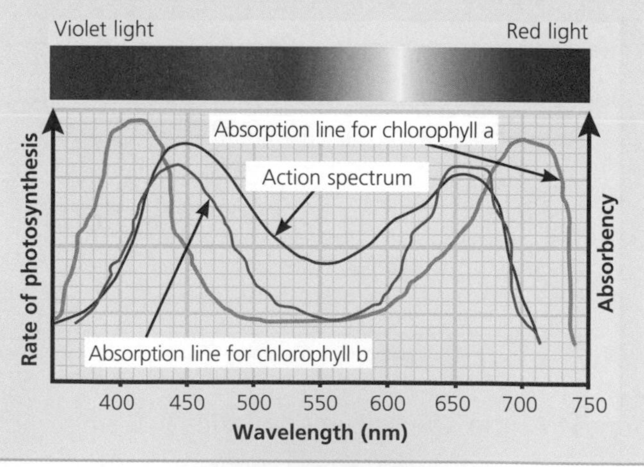

## Leaves

The first diagram below shows a section of a typical leaf. The structure is made up of four distinct layers: the **upper epidermis**, the **palisade layer**, the spongy **mesophyll** and the **lower epidermis**.

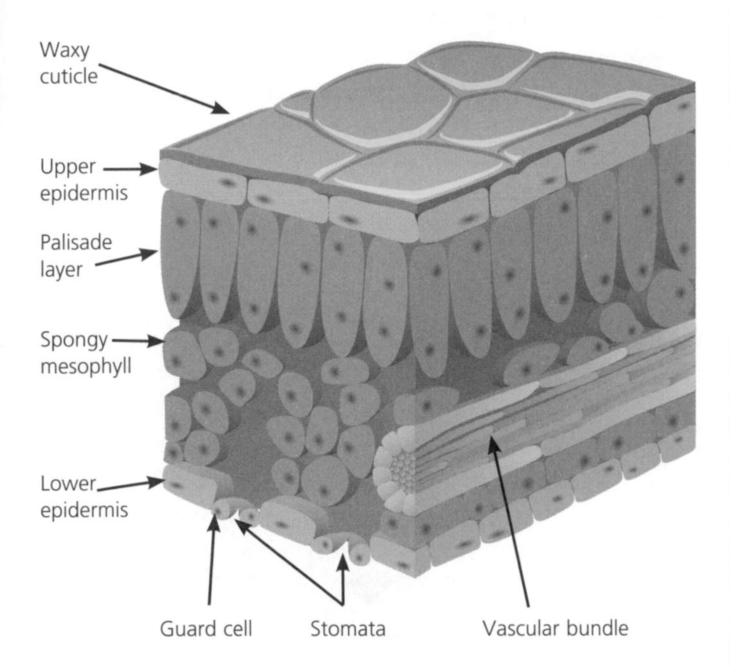


The leaf palisade cell is adapted to its function by having a large concentration of chloroplasts.

**A Palisade Cell**

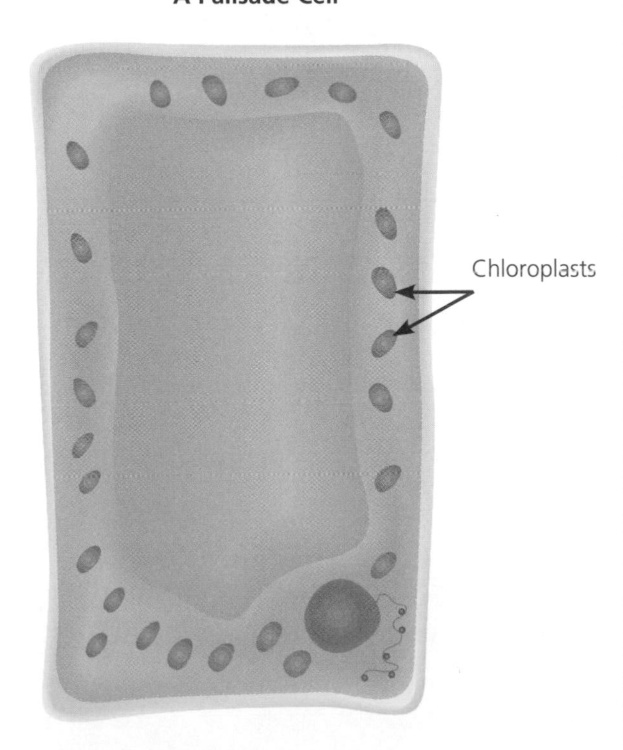


HT The functions of the parts of the leaf shown in the diagram (opposite) are explained below:

- The waxy cuticle restricts water loss from the leaf.
- The upper epidermis is transparent to allow sunlight through to the layer below.
- The cells in the palisade layer are positioned near the top of the leaf to absorb the maximum amount of light and are packed with chloroplasts, which absorb the light energy needed for photosynthesis.
- The spongy mesophyll contains lots of air spaces connected to the stomata to ensure the optimum exchange of gases.
- The lower epidermis contains most of the stomata.
- Stomata allow the diffusion of gases (i.e. carbon dioxide and oxygen) in and out of the leaf.
- Guard cells control the size of the stomata to restrict water loss.
- Vascular bundles contain xylem and phloem to transport water and sugars through the leaf.
- Leaves also have a large internal surface area-to-volume ratio, which improves the rate of gaseous exchange between photosynthesising cells and the air spaces.

# Diffusion and Osmosis

## Diffusion

The cell membrane controls which substances enter and leave the cell. Living cells need to obtain oxygen, glucose, water and minerals from their surroundings and get rid of waste products, such as carbon dioxide. These substances pass through the cell membrane by **diffusion** – the movement of a substance **from a region of high concentration to a region of low concentration**.

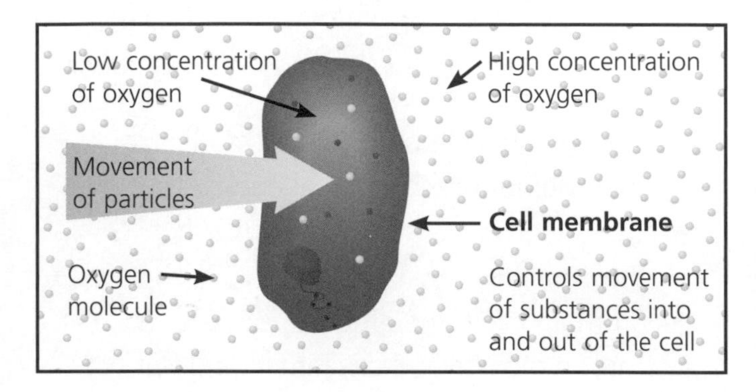


There is a low concentration of oxygen inside the cell because supplies are constantly being used up in respiration; there is a high concentration outside the cell because oxygen is constantly being replaced.

Particles have **random movement**. Diffusion is the net (overall) movement of particles from high concentration to low concentration.

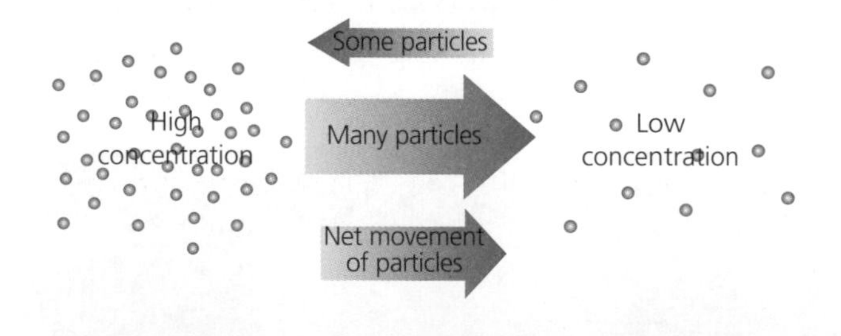


HT The rate of diffusion is increased when:
- the cell membrane's surface area is bigger
- there is a bigger difference between concentrations (a steeper **concentration gradient**)
- the particles have a shorter distance to travel.

## Osmosis

**Osmosis** is the diffusion of water from a dilute solution to a more concentrated solution through a **partially permeable membrane.**

Water moves from a cell with a high concentration of water to a cell with a lower concentration of water (diagram 1) until there is the same concentration of water in each cell (diagram 2).

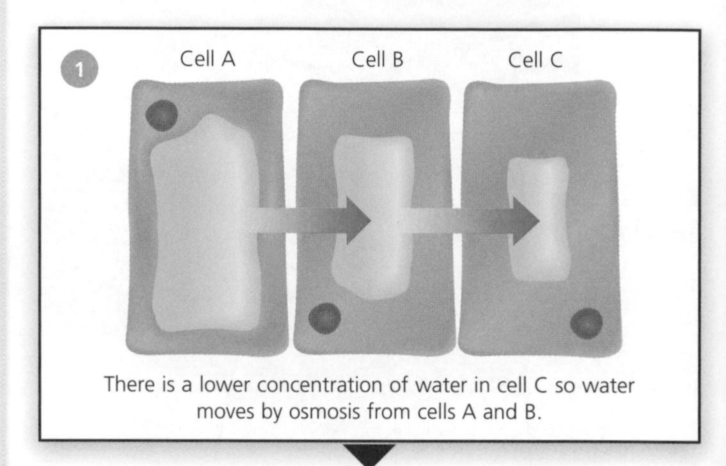


There is a lower concentration of water in cell C so water moves by osmosis from cells A and B.

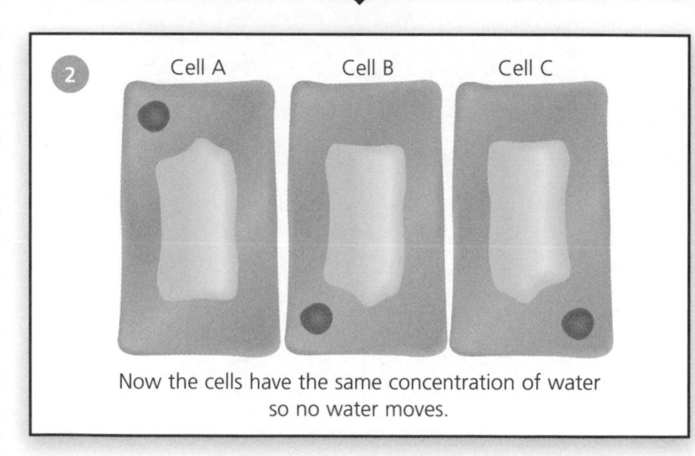


Now the cells have the same concentration of water so no water moves.

The cell walls do not affect the movement of water or dissolved substances because they are freely permeable. The cell membranes allow the movement of water but restrict the movement of dissolved substances.

Remember:
- **osmosis** is the movement of **water** molecules
- **diffusion** is the movement of **other substances**.

## HT More on Osmosis

When predicting the direction of water movement, the only thing that matters is the concentration of the water because the solute molecules cannot pass through the membrane; only the water molecules can.

The water particles move randomly, colliding with each other and passing through the membrane in both directions. Overall, more particles move from the area where water is at high concentration to low concentration. The effect is to gradually dilute the solution.

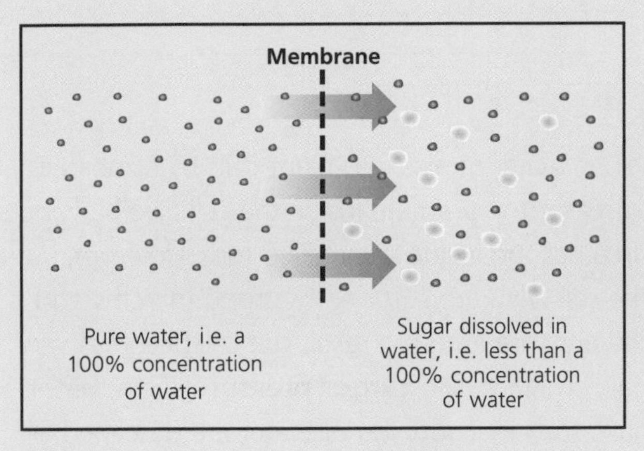


At the root hair cells, water gradually moves from the soil into the cell by osmosis along a concentration gradient.

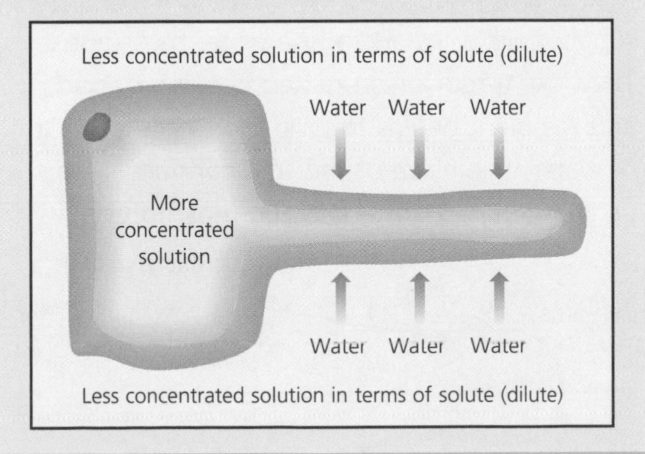


## Water in Plants

Plants use water to:
- keep their leaves cool
- transport minerals
- enable them to photosynthesise to produce glucose
- keep cells firm and so keep the plant rigid.

Plants need to balance the amount of water they take in with the amount they lose. Water is taken in via the roots, which are specially adapted to increase water intake by having root hair cells. These cells increase the surface area of the root, making a greater area for absorption.

The water then travels through the plant to the leaves, along a concentration gradient from an area of high concentration of water to an area of low concentration.

When water reaches the leaves it can be lost by **evaporation** (**transpiration**) through stomata. There are two adaptations that can reduce the rate at which water is lost from the leaves into the atmosphere:
- having a waxy cuticle on the surface of the leaf
- having the majority of the stomata on the lower surface of the leaf, where they are less exposed to sunlight.

## Maintaining Support

Plant cells have **inelastic cell walls** which, together with the water inside the cells, are essential for the support of young non-woody plants. The cell wall prevents cells from bursting due to excess water entering them and, at the same time, contributes to turgidity.

A lack of water can cause plants to wilt. As the amount of water inside the cells reduces, the cells become less rigid due to reduced pressure.

If there is plenty of water in the soil, stomata open to allow transpiration and the diffusion of gases. The plant cells will be full of water so the plant stays erect. This is the main method of support for young plants.

When there is not enough water available in the soil, stomata close to try to prevent transpiration and the diffusion of gases, so rate of photosynthesis has to be reduced. The plant cells are short of water and lose their turgor pressure, which causes the plant to wilt.

## Osmosis in Animal Cells

Water diffuses in and out of animal cells by **osmosis**. But animal cells do not have a cell wall, so if too much water enters a cell, it could burst.

| Ideal Shape | Swollen | Shrivelled |
|---|---|---|
| When red blood cells are in solutions that have the same concentration as their cytoplasm they retain their shape. | When in a weaker solution, they absorb water and swell up, and they may burst. | When in a more concentrated solution, they lose water and shrivel up. |

### More on Osmosis in Animal Cells

The cytoplasm of a red blood cell has the same water concentration as plasma. There is no net movement of water into or out of the cell, so it can maintain its biconcave shape.

If blood cells are put into pure water, they gain lots of water by osmosis. There is nothing to prevent the water entering the cell because, unlike plant cells, there is no inelastic cell wall to resist the outward pressure of the water so they eventually burst. This is called **lysis**.

Blood cells in a concentrated solution (very little water) lose water by osmosis. They shrivel up and become **crenated** (i.e. they have scalloped edges).

Animal and plant cells respond differently to osmosis because animal cells do not have a cell wall. When plant cells absorb water they expand, but resistance from the cell wall eventually stops them from absorbing any more water and bursting. Animal cells do not have a cell wall so they absorb more and more water until lysis occurs.

### Turgor Pressure

When water moves into plant cells by osmosis it increases the pressure inside the cell (like blowing up a balloon inside a cereal carton). However, the cell walls are sufficiently strong to withstand the pressure and as a result the cell becomes very rigid. This is called **turgor pressure**. Cells that have sufficient supplies of water are described as maintaining their turgor, i.e. staying rigid.

When all the cells are fully **turgid**, the plant is rigid and upright. However, if water is in short supply, cells will start to lose water by osmosis. They lose turgor pressure and become **flaccid**, and the plant begins to wilt. If the cells lose a lot of water, the inside of the cells contract like a deflating balloon. This is called **plasmolysis**.

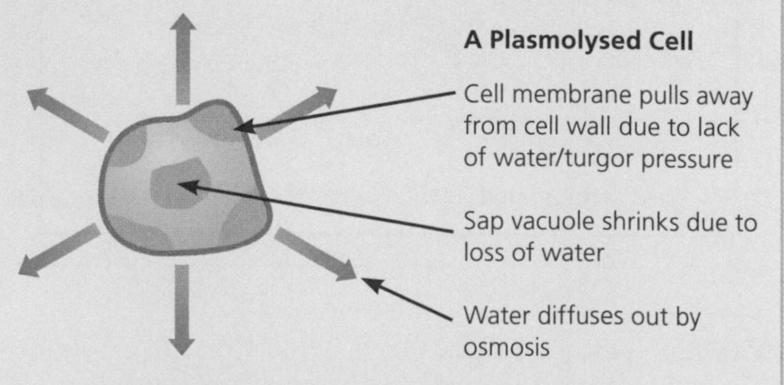


In practice, plant cells will only become plasmolysed if they are exposed to high solute concentrations, e.g. salt.

## Water Loss from Leaves

**Transpiration** and **water loss** are an unavoidable consequence of **photosynthesis**. A plant therefore needs to balance its water loss with its water uptake to avoid wilting – or even death. The leaf consists of moist cells exposed to the atmosphere via the **stomata**, which can be found on the undersurface of the leaf. A lot of stomata are needed because they allow the exchange of gases during photosynthesis, but they also allow water molecules to pass out of the leaf.

HT The number, position, size and distribution of stomata vary between plants, depending on the environment they live in and, therefore, the amount of water they require.

The leaf is adapted to be able to reduce water loss. The turgidity of guard cells changes in relation to the light intensity and availability of water in order to regulate the size of the stomatal openings.

When a plant is exposed to sunlight and a sufficient supply of water, photosynthesis occurs in the cells of its leaves, including the guard cells. Photosynthesis produces glucose and this creates a concentration gradient so that water enters the leaves by osmosis. This makes the guard cells turgid and the stomata open fully allowing the exchange of gases and water loss.

## The Structure of a Plant

The **flower** contains the reproductive organs of the plant, which are required to make seeds.

**Leaves** are broad, thin and flat to provide a large surface area to absorb sunlight, which is needed for photosynthesis.

In most plants, the **stem** supports the plant and transports substances from the roots to the leaves by the transport tissues: **xylem** and **phloem**. Xylem transport water and soluble mineral salts from the root hair cells to the leaves (**transpiration**) to replace water lost by evaporation and photosynthesis. Phloem tubes allow the movement of food substances around

the plant (**translocation**). For example, sucrose is transported to storage organs (usually in the roots) and converted to starch. Or it may be sent to growing regions such as meristems or buds.

The xylem and phloem form a continuous system of tubes from roots to leaves. The diagram below shows how xylem and phloem are arranged in the plant:

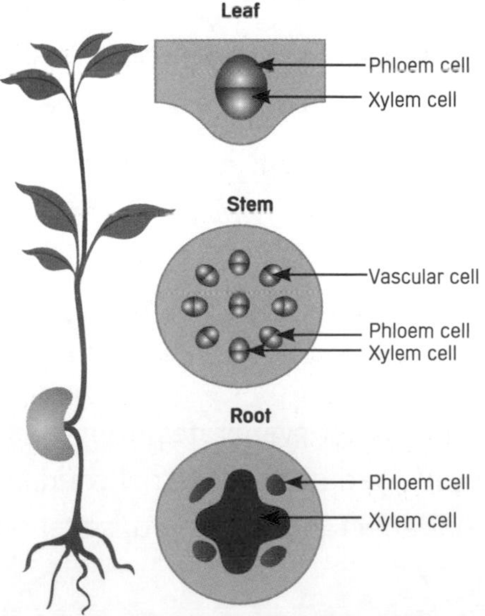

## Transpiration

Transpiration is the movement of water through the whole plant. The transpiration stream is powered by the evaporation of water from the leaf:

- Water evaporates from the internal leaf cells through the stomata.
- Water passes by osmosis from the xylem vessels to leaf cells, which pulls the entire thread of water in that vessel upwards by a very small amount.
- Water enters the xylem from root tissue to replace water which has moved upwards.
- Water enters root hair cells by osmosis to replace water which has entered the xylem.

Minerals from the soil are absorbed and carried as dissolved ions (e.g. $K^+$, $NO_3^-$) in the transpiration stream.

HT Xylem vessels are hollow tubes made from **dead** plant cells. This hollow centre is called a **lumen**, much like the centre of blood vessels. The cellulose cell walls are thickened and strengthened with a waterproof substance.

Phloem cells are long columns of **living** cells, able to translocate (move) sugars both up and down the plant.

**Roots** anchor the plant in the ground and absorb water and minerals. The root hair cells have an enormous surface area for the absorption of water.

To sum up, transpiration provides plants with water for:

- cooling (via evaporation from the leaves)
- photosynthesis
- support
- transport of minerals.

## Factors Affecting the Rate of Transpiration

The rate that water evaporates from the leaves of a plant is affected by the external conditions. The factors that affect the rate of transpiration are:

- **Light** – more light increases the rate of photosynthesis and therefore also increases the transpiration rate.
- **Air movement** (wind) – transpiration increases as the movement of the air increases.
- **Temperature** – the rate of photosynthesis increases as temperature increases, and this increases the rate of transpiration as a result.
- **Humidity** – humidity is the amount of water vapour in the air. Increased humidity decreases the rate of transpiration.

HT **How Water Vapour Exits the Leaf**

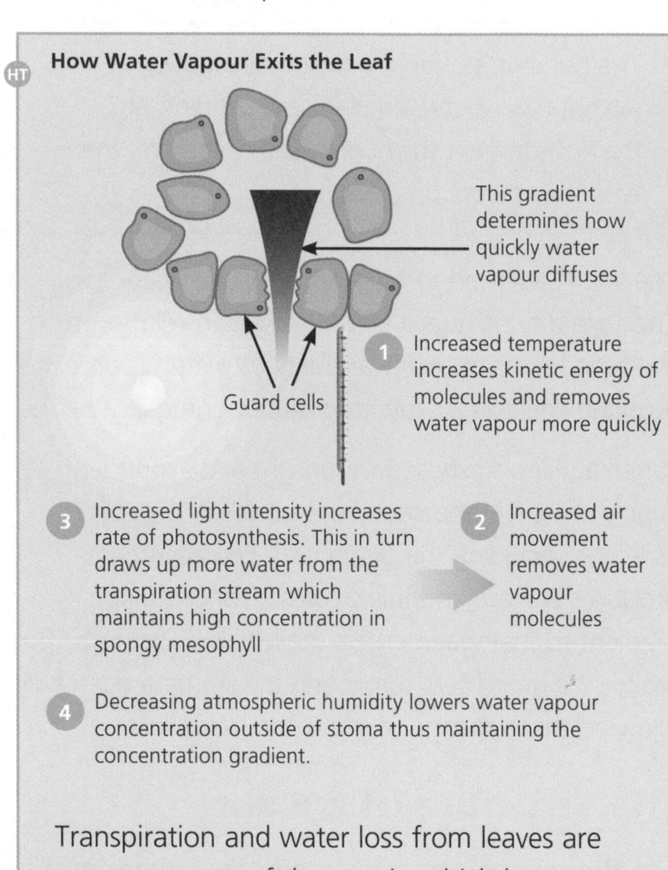


Transpiration and water loss from leaves are a consequence of the way in which leaves are adapted to efficient photosynthesis.

When conditions on the outside of the leaf change they affect the diffusion gradient of water between the atmosphere and the spongy mesophyll. The diagram above shows how the gradient is affected.

The cellular structure of a leaf is adapted to reduce water loss through changes in guard cell turgidity.

Where stomata are placed can also affect transpiration rate. Plants that live in arid (dry) climates often have fewer stomata overall and have no stomata on the upper surface of the leaf because heat from the Sun will increase evaporation here.

Leaves are also adapted to reduce excessive water loss:

- they have a waxy cuticle.
- the number, position, size and distribution of stomata vary between plants depending on their environment (which affects how much water they need).

## Measuring Rate of Transpiration

A leafy shoot's rate of transpiration can be measured using a **potometer** (see diagram below).

The shoot is held in a test tube with a bung around the top to prevent any water from evaporating (this would give a false measurement of the water lost by transpiration).

As the plant transpires, it takes up water from the test tube to replace that which it has lost. All the water is then pulled up, moving the air bubble along.

The distance the air bubble moves can be used to calculate the plant's rate of transpiration for a given time period.

The experiment can be repeated, varying a different factor each time, to see how each factor affects the rate of transpiration.

HT When light intensity is high and photosynthesis is proceeding at a rapid rate then the sugar concentration rises in photosynthesising cells, e.g. palisade and guard cells. Guard cells can respond to this by increasing the rate of water movement in the transpiration stream. This in turn provides more water for photosynthesis.

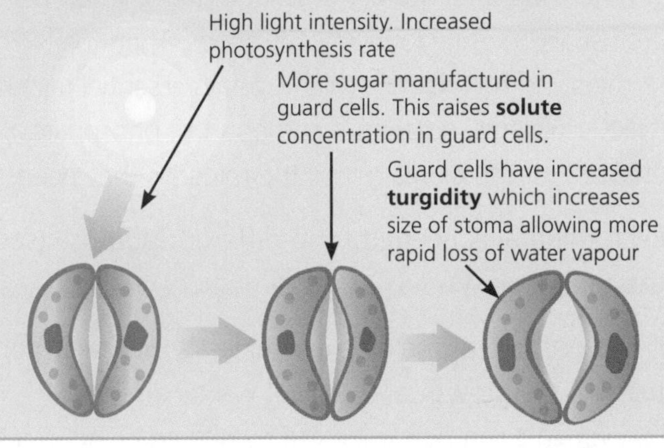


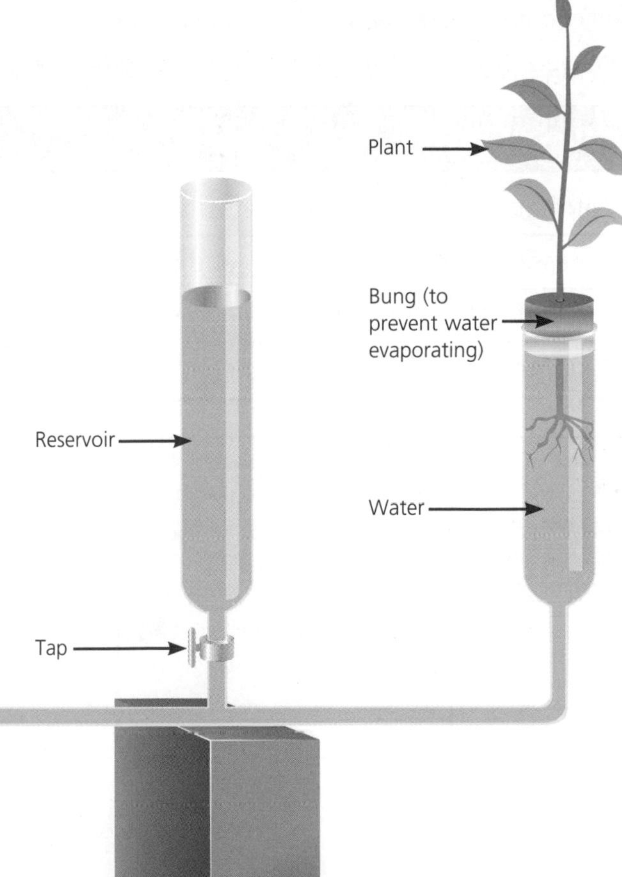


**A potometer**

# B4 Plants Need Minerals

## Essential Minerals

Essential **minerals** – nitrates, potassium, phosphates and magnesium – are needed to make proteins and other compounds to keep the plant healthy and growing properly. Plants absorb these essential minerals through their root hairs as ions dissolved in water.

The minerals are naturally present in the soil but they are usually in quite low concentrations. Therefore, farmers use **fertilisers**, which contain essential minerals (**NPK** – nitrates; potassium compounds; phosphates), to ensure that the plants get all the minerals they need.

Each mineral is needed for a different purpose (see the table below).

If one or more of the essential minerals is missing from the soil, the growth of the plant will be affected. Experiments can be carried out by placing plants in a soil-less culture, where each trial has one essential mineral removed. The results are shown in the diagrams below.

## HT Active Transport

Substances are sometimes absorbed **against a concentration gradient**, i.e. from a low to a high concentration. This is the opposite direction to which normal diffusion occurs. This process is called **active transport** and it requires energy from respiration in the same way that pulling a trolley up a hill would require energy. Plants absorb ions from very dilute solutions.

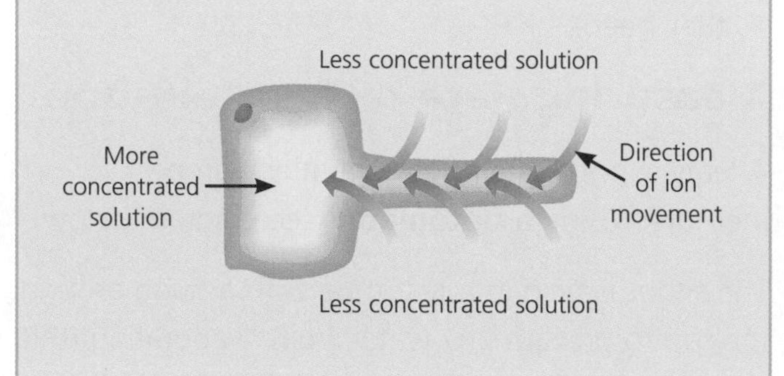


| Mineral | Why it is Needed | How it is Used |
|---|---|---|
| Nitrates | To make proteins for cell growth. | HT To provide nitrogen for amino acids that form the proteins. |
| Potassium | For respiration and photosynthesis. | HT To help the enzymes in these processes. |
| Phosphates | For respiration and growth. | HT To provide phosphorus for DNA and cell membranes. |
| Magnesium | For photosynthesis. | HT To make the chlorophyll molecule for photosynthesis. |

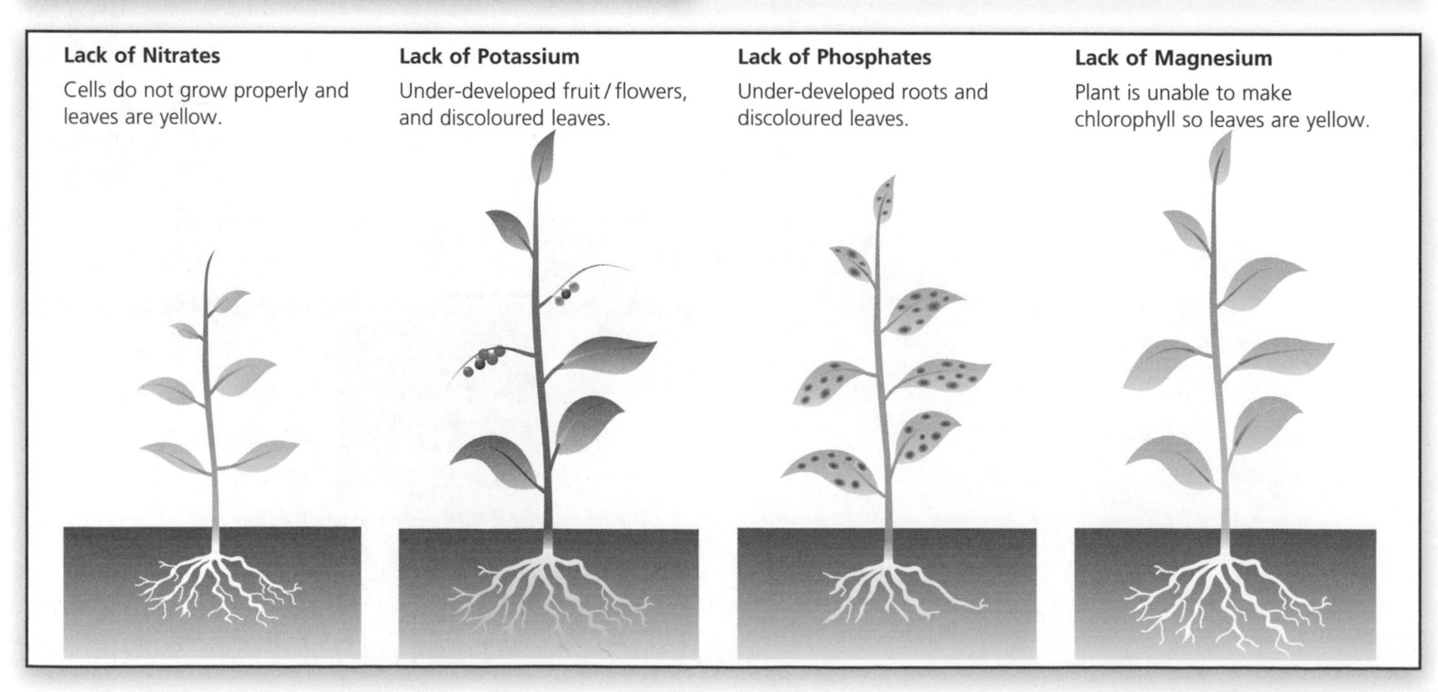

## Decay

**Decay** is a process involving the breakdown of complex substances into simpler ones by microorganisms. The rate at which microorganisms break down substances is affected by several factors:

- **Temperature** – microorganisms responsible for decay work at their optimum at 40°C. Below this temperature, the microorganisms work slower; above this temperature the enzymes they contain are denatured, which means they cannot carry out reactions.
- **Amount of oxygen** – microorganisms' rate of activity increases as the amount of oxygen in the air increases. So, generally, the more oxygen there is the better.
- **Amount of water** – microorganisms prefer moist conditions. However, if there is too much water, the air spaces in the substrate become waterlogged, reducing the supply of oxygen. This in turn slows down the decay rate.

Decay is important for plant growth because it releases minerals into the soil for their healthy growth.

When food and organic material is exposed to air, moisture and warmth, microorganisms will quickly grow on it. Storing food in dry, cold conditions prevents microorganisms and fungi from growing.

HT The ideal conditions for the microorganisms that cause decay are those that increase their respiration, growth and reproduction rates.

**Temperature** – as the temperature is increased, the microorganisms' rate of respiration and growth increase until it reaches 40°C, which is the optimum temperature (i.e. at this temperature the microorganisms' growth and respiration rates are as high as they can be). Above 40°C the enzymes are denatured, so decay stops.

**Amount of oxygen** – increasing the amount of oxygen in the air increases the microorganisms' rate of respiration, which means they release more energy enabling them to grow and reproduce more quickly. There is no optimum amount of oxygen – the more there is, the better.

**Amount of water** – microorganisms grow quickest in moist conditions, which increases the rate of decay. Having too much or too little water present slows down their growth and, therefore, the rate of decay.

Bacteria and fungi are **saprophytes** – they feed on dead organic material by secreting enzymes externally onto their food and then absorbing the digested products. This is called **extra-cellular digestion**. Saprophytes are essential for decay.

## Food Preservation

Microorganisms will feed on any source of food and cause it to decay if the conditions are right. Food can be preserved (by removing the oxygen, heat or moisture that the microorganisms need) using one of the following methods:

- **Canning** – food can be heated to destroy microbes and sealed inside sterile cans or bottles to prevent microorganisms from getting any oxygen.
- **Refrigeration and freezing** – food can be kept at low temperatures to slow down the growth and reproduction of microorganisms.
- **Pickling** – food can be pickled in vinegar; its low pH slows down the growth of microorganisms by denaturing the microorganisms' enzymes.
- **Adding sugar** – adding sugar (or salt) to a food makes the conditions too concentrated for the microorganisms to survive. The higher solute concentration draws water out of microorganisms' cells by osmosis, which kills them.
- **Drying** – drying food reduces the moisture available to microorganisms, preventing them from growing.

# B4 Decay

## Food Decay

### Example

An experiment can be carried out to show that decay is caused by external factors – such as the presence of microorganisms and air (oxygen) (see diagram alongside). The following method can be used:

1. Pour a solution containing nutrients into Flask A. Melt and shape the neck of the flask.
2. Boil the nutrient solution to kill microorganisms and drive out air.
3. Seal the neck of the flask.
4. Pour more of the same nutrient solution into another flask (Flask B), but this time snap off the neck.

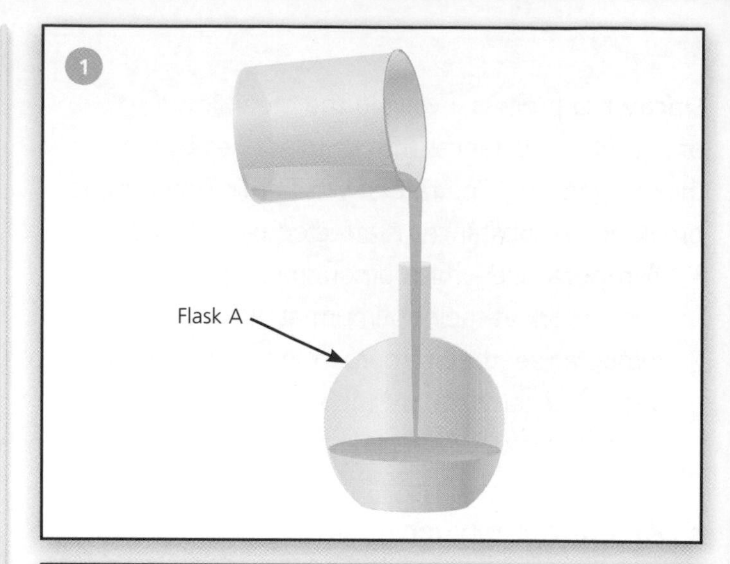


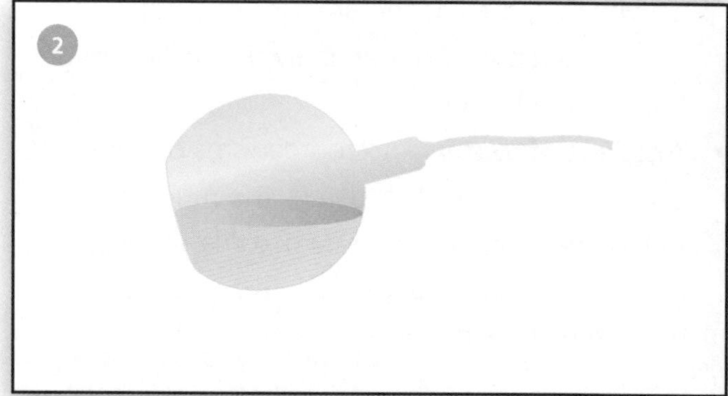


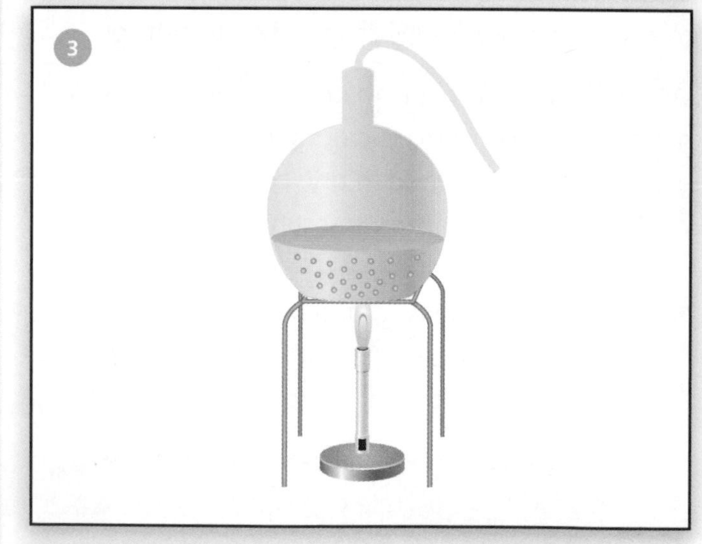


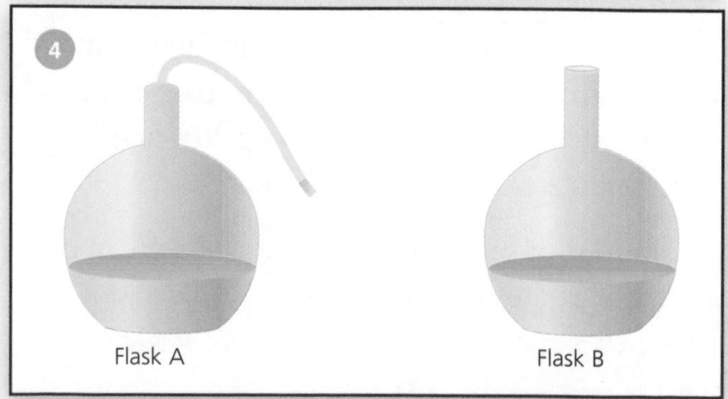


The solution that is in the flask that had the neck snapped off (Flask B) will start to decay within days because microorganisms and oxygen will be able to enter the flask, but the solution in the other flask (Flask A) will show no signs of decay as long as it remains sealed.

## Decomposers

When dead organisms or waste materials decay, minerals are released which can then be re-used by other living organisms, particularly plants.

Various soil organisms, including bacteria, fungi, earthworms, maggots and woodlice, help with the process of decay.

Earthworms, woodlice and maggots are known as **detritivores**; they feed on dead organisms and the waste (detritus) produced by living organisms. Detritivores speed up the process of decay because they break down detritus into small particles which have a large surface area, making it easier for decomposers to feed on. The faeces of detritivores also provide food for decomposers like bacteria and fungi.

Decomposer organisms (microorganisms) are used by humans to break down waste:

- They feed on human waste in sewage treatment works.
- They break down plant waste in compost heaps.

## Intensive Farming

Farmers are increasingly faced with the challenge of producing higher yields of food at lower costs. To achieve this, they can use **intensive farming** practices to produce as much food as possible from the available land, plants and animals. Pesticides, insecticides and herbicides are used in this process.

Intensive farming practices are efficient and include keeping animals in carefully controlled environments where their temperature and movement are limited. For example, fish farms breed huge numbers of fish in enclosed nets in the sea or lakes, and battery farms raise chickens in cages. However, this can raise ethical dilemmas – some people find this morally unacceptable because the animals have a poor quality of life.

**Pesticides** are used to kill pests that can damage crops or farm animals. Great care needs to be taken with pesticides because they can harm other organisms (non-pests) and build up (accumulate) in food chains, harming animals at the top levels.

**Insecticides** are used to kill insect pests, **fungicides** are used to kill fungi and **herbicides** are used to kill weeds that would compete with the crop for water and nutrients.

## Accumulation of Pesticides

Herbicides and pesticides reduce the amount of weeds and pests and, therefore, the energy that is taken by competing plants and pests. However, they can cause problems. Pesticides can flow into rivers where they can be absorbed by **algae** (tiny plants). The algae is eaten by small aquatic organisms, which in turn are eaten by larger organisms.

The pesticide increases in concentration along the food chain because the organisms cannot excrete the chemical – this is called **persistence** and means that virtually all of the pesticide is passed on to the next trophic level.

So, while the pesticide may not damage the algae or the smaller organisms, the effect of accumulation could easily kill the larger organisms.

The accumulation of pesticides can be shown by the following example.

1. Each tiny plant absorbs 2 units of pesticide.
2. A small fish eats five plants – it has eaten $5 \times 2 = 10$ units of pesticide.
3. A bird eats four fish – it has eaten $4 \times 10 = 40$ units of pesticide.

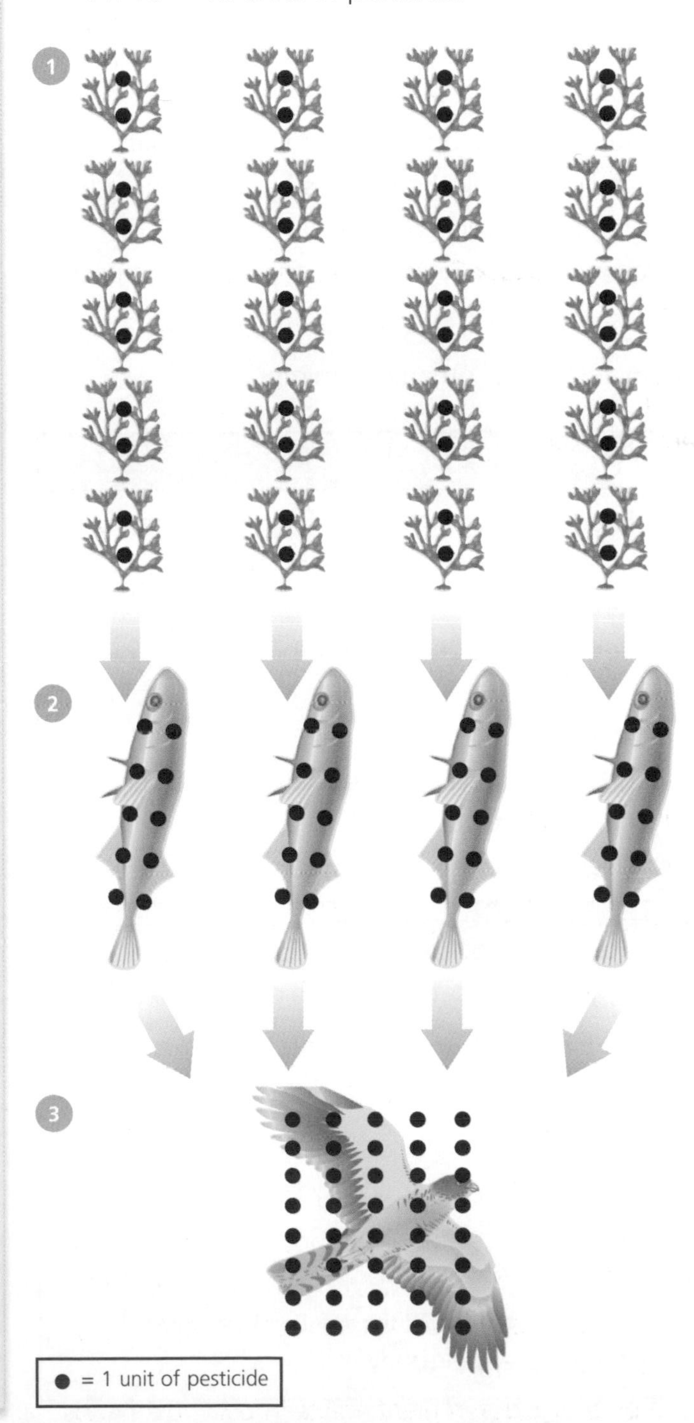

## Organic Farming

Some people have become very concerned about the impact of intensive farming on the quality of food produced, the welfare of animals and the environment.

**Organic** farmers ensure that the quality of their food is as high as possible, while maintaining the welfare of their animals and minimising the impact on the environment by:

- using animal manure or compost instead of chemical fertilisers
- growing nitrogen-fixing crops such as peas or clover to trap nitrogen in the soil
- rotating their crops to maintain soil fertility
- avoiding chemical pesticides by weeding and using biological controls
- varying seed planting times to discourage pests.

The table below gives some of the advantages and disadvantages of organic farming:

| Advantages |
|---|
| • Food crops and the environment are not contaminated with artificial fertilisers or pesticides.<br>• Soil fertility is maintained through the use of organic fertilisers, and soil erosion is limited.<br>• Biodiversity in the local environment is promoted because hedgerows and other habitats are conserved.<br>• Livestock have space to roam. |
| **Disadvantages** |
| • Organic fertiliser takes time to rot and does not supply a specific balance of minerals.<br>• Weeds have to be removed by hand.<br>HT • Organic farming is less efficient because some crops and the energy they contain are lost to pests and diseases.<br>• Free-roaming livestock use more energy and lose more body heat. This requires more food to increase their body mass. |

## Hydroponics

A relatively new development in intensive farming is **hydroponics**. This is the term given to growing plants without soil. The plants grow with their roots in a solution. Minerals needed for healthy plant growth are added to the solution. Plants can be grown in glasshouses without soil using this system and it is also useful in areas where the soil is very thin or barren.

Certain plants, e.g. tomatoes, can be grown hydroponically in greenhouses. The temperature can be controlled using heaters; the light intensity can be controlled using lamps; and the carbon dioxide concentration can be controlled using chemicals; or as a by-product of fossil fuel heaters.

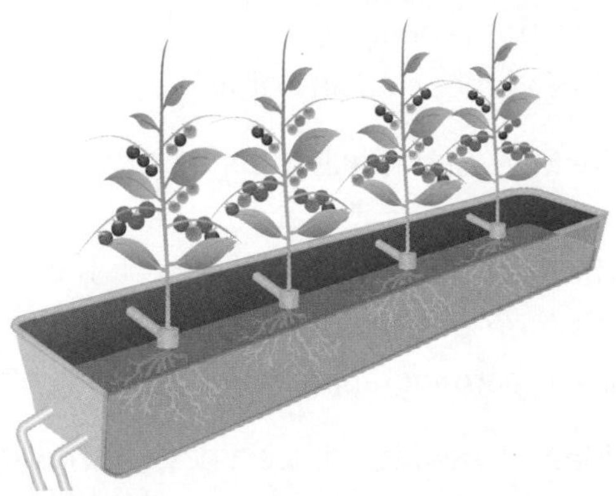

HT The table below gives some of the advantages and disadvantages of hydroponics:

| Advantages |
|---|
| • The mineral levels added to the solution can be controlled carefully and adjusted to the type of plant.<br>• There is a reduced risk of the plants becoming diseased. |
| **Disadvantages** |
| • The plants have to be supported because they have no anchorage for their roots.<br>• Expensive fertilisers are a compulsory aspect of the system to supply the plant with minerals. |

## Biological Control

Instead of using pesticides, some farmers prefer to introduce a **predator** to reduce the number of pests. This is called **biological control**. For example, the cottony cushion scale (an insect) was a pest that attacked citrus fruit crops in America. However, when farmers introduced the ladybird beetle, the pest's numbers were significantly reduced.

Another example is the prickly pear cactus. It grew at a very fast rate and was taking over a lot of useful farm land, so a moth whose larva fed on the cactus' tissues was introduced, which reduced the number of cacti.

However, it is important to remember that whether biological control or pesticides are used to get rid of pests, thought must be given to the effect on the rest of the organisms in the food chain or web. In the example below, suppose the slug population was controlled by introducing a natural predator. The new organism would compete with mistle thrushes and potentially reduce their number.

The table below lists the advantages and disadvantages of biological control:

| Advantages |
|---|
| • There is no need for chemical pesticides.<br>• Once a predator is introduced, it can have an impact over many years. There is no need to repeat the treatment.<br>• The pest cannot become resistant to the predator (as it can to pesticides). |
| **Disadvantages** |
| • The pest is not eliminated; it only has its numbers reduced.<br>• The predator may not eat the pest.<br>• The predator may eat useful species.<br>• The predator's population may grow out of control (e.g. the cane toad in Australia) and have unanticipated effects.<br>• The predator may move beyond the area in which it was introduced and become a nuisance. |

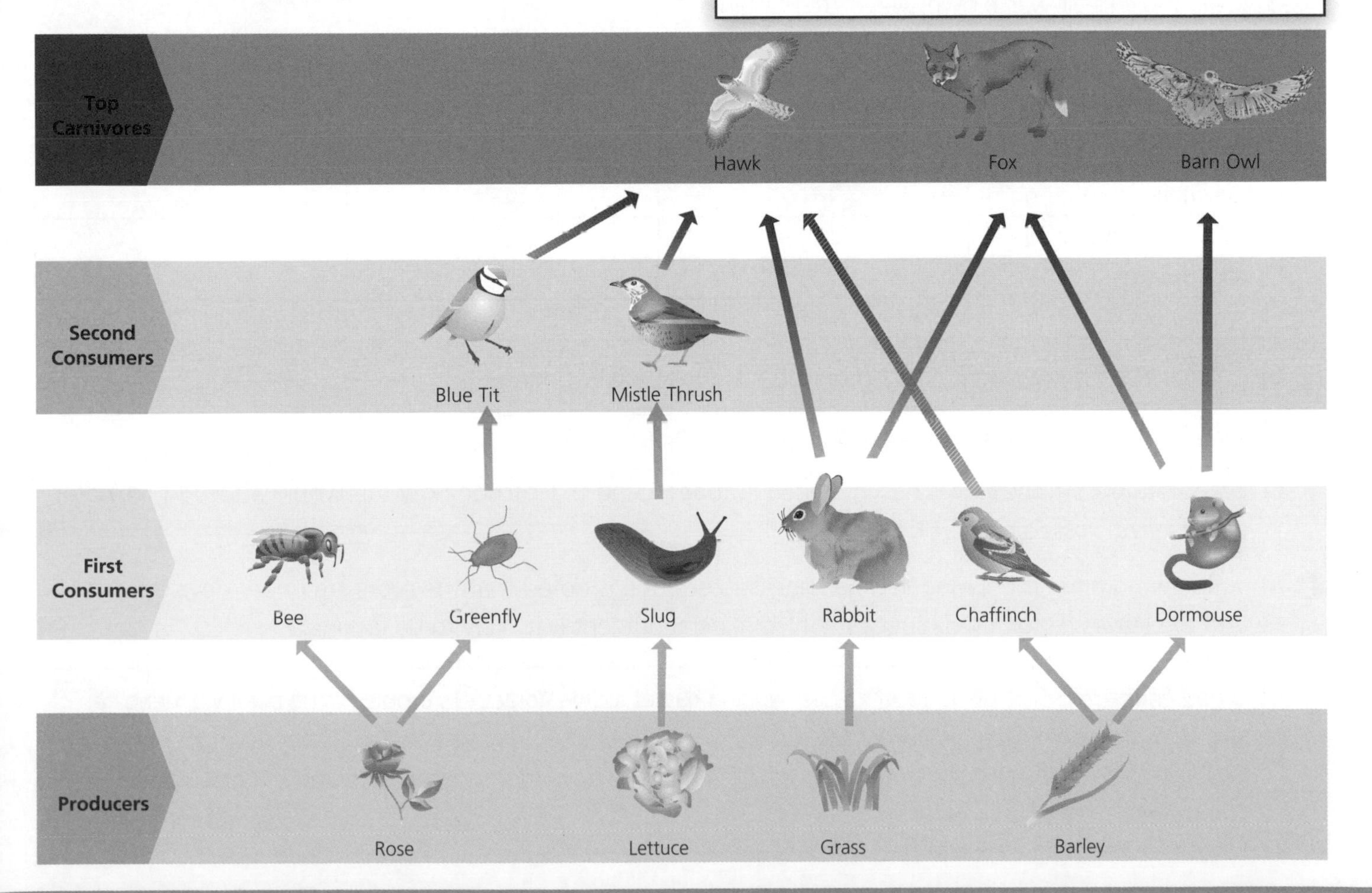

# B4 Exam Practice Questions

1. This diagram shows an aquatic plant called elodea.

   **a)** It obtains its food through the process of photosynthesis. Name two raw materials the plant needs to carry out this process. **[2]**

   **b)** The food it manufactures is in the form of glucose. A scientist traced the carbon in the glucose using a radioactive form of the element. He found radioactivity in other compounds apart from glucose. Suggest two compounds that might have been radioactive apart from the glucose. **[2]**

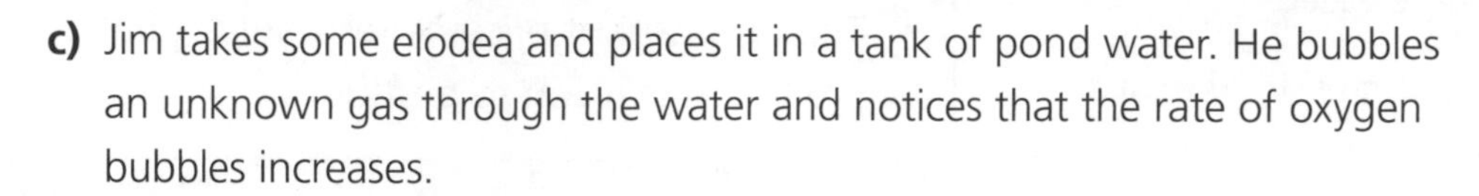

   **c)** Jim takes some elodea and places it in a tank of pond water. He bubbles an unknown gas through the water and notices that the rate of oxygen bubbles increases.

   **i)** Suggest what the unknown gas might be. **[1]**

   **ii)** Jim increases the concentration of the unknown gas tenfold. Describe the pattern of bubble production which would result from this change. **[1]**

2. In an experiment, a plant biologist set up an investigation to measure the rate of transpiration in a privet shoot. She set up tubes like the one in the diagram, weighed them and exposed them to different conditions as follows:

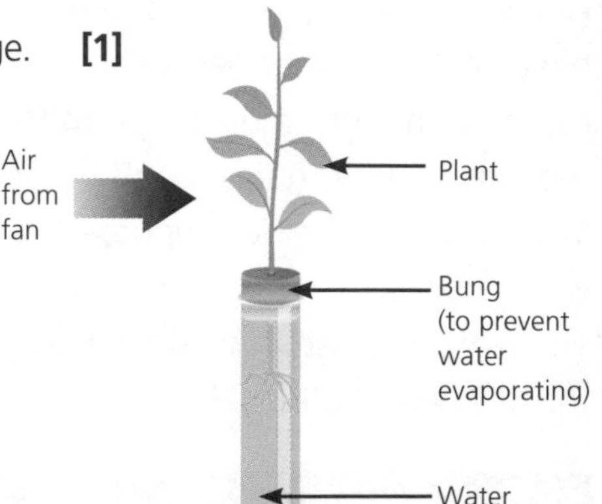


   **A** - Left to stand in a rack
   **B** - Cold moving air was blown over it from a fan
   **C** - A radiant heater was placed next to it

   Each tube was left for 6 hours and then re-weighed. She recorded the masses in this table:

| Tube | A | B | C |
|---|---|---|---|
| Mass at start (g) | 41 | 43 | 42 |
| Mass after 6 hours (g) | 39 | 35 | 37 |
| Mass loss (g) | 2 | 8 | 5 |
| % mass loss | 4.9 | | 11.9 |

   **a)** Calculate the percentage mass loss in tube B. Show your working. **[2]**

   **b)** Which factor increased the rate of transpiration the most? **[1]**

   **c)** Evaporation from the leaves has increased in tubes B and C. Describe how this would affect liquid water in the xylem vessels of the plant. **[1]**

3. **a)** A flowering plant is growing in a hanging basket but it is noted that the roots are under-developed and the leaves discoloured. Suggest which mineral nutrient is it likely to be lacking. **[1]**

   HT **b)** A gardener adds some plant 'food' to the hanging basket to counteract this problem. The plant will need to absorb the mineral ions in the plant food through its roots. The plant also absorbs water through its roots. Explain, using ideas about osmosis and active transport, how the water uptake and mineral uptake will differ. **[4]**

# Fundamental Chemical Concepts

You need to have a good understanding of the concepts (ideas) on the next four pages, so make sure you revise this section before each exam.

## Elements and Compounds

An **element** is a substance made up of just one type of **atom**. Each element is represented by a different chemical symbol, for example:

- Fe represents iron
- Na represents sodium.

Atoms have a positive nucleus orbited by negative electrons.

These elements (and their chemical symbols) are all arranged in the **Periodic Table** (see below).

**Compounds** are substances formed from the atoms of two or more elements, which have been joined together by one or more chemical bonds, for example, $H_2O$, $CaCO_3$ and $C_6H_{12}O_6$.

**Ions** are atoms or small molecules that have a charge, for example, $Na^+$, $Cl^-$, $NH_4^+$ and $SO_4^{2-}$.

A positive ion is formed when an atom loses electrons; a negative ion is formed when an atom gains electrons.

**Covalent bonds** are formed when two atoms share a pair of electrons. (The atoms in molecules are held together by covalent bonds.)

**Ionic bonds** are formed when atoms lose or gain electrons to become charged ions; the positive ions attract the negative ions.

**The Periodic Table**

| 1 | 2 | | | | | | | | | | | 3 | 4 | 5 | 6 | 7 | 0 |
|---|---|---|---|---|---|---|---|---|---|---|---|---|---|---|---|---|---|
| | | | | | | | | 1 **H** hydrogen 1 | | | | | | | | | 4 **He** helium 2 |
| 7 **Li** lithium 3 | 9 **Be** beryllium 4 | | | | | | | | | | | 11 **B** boron 5 | 12 **C** carbon 6 | 14 **N** nitrogen 7 | 16 **O** oxygen 8 | 19 **F** fluorine 9 | 20 **Ne** neon 10 |
| 23 **Na** sodium 11 | 24 **Mg** magnesium 12 | | | | | | | | | | | 27 **Al** aluminium 13 | 28 **Si** silicon 14 | 31 **P** phosphorus 15 | 32 **S** sulfur 16 | 35.5 **Cl** chlorine 17 | 40 **Ar** argon 18 |
| 39 **K** potassium 19 | 40 **Ca** calcium 20 | 45 **Sc** scandium 21 | 48 **Ti** titanium 22 | 51 **V** vanadium 23 | 52 **Cr** chromium 24 | 55 **Mn** manganese 25 | 56 **Fe** iron 26 | 59 **Co** cobalt 27 | 59 **Ni** nickel 28 | 63.5 **Cu** copper 29 | 65 **Zn** zinc 30 | 70 **Ga** gallium 31 | 73 **Ge** germanium 32 | 75 **As** arsenic 33 | 79 **Se** selenium 34 | 80 **Br** bromine 35 | 84 **Kr** krypton 36 |
| 85 **Rb** rubidium 37 | 88 **Sr** strontium 38 | 89 **Y** yttrium 39 | 91 **Zr** zirconium 40 | 93 **Nb** niobium 41 | 96 **Mo** molybdenum 42 | [98] **Tc** technetium 43 | 101 **Ru** ruthenium 44 | 103 **Rh** rhodium 45 | 106 **Pd** palladium 46 | 108 **Ag** silver 47 | 112 **Cd** cadmium 48 | 115 **In** indium 49 | 119 **Sn** tin 50 | 122 **Sb** antimony 51 | 128 **Te** tellurium 52 | 127 **I** iodine 53 | 131 **Xe** xenon 54 |
| 133 **Cs** caesium 55 | 137 **Ba** barium 56 | 139 **La*** lanthanum 57 | 178 **Hf** hafnium 72 | 181 **Ta** tantalum 73 | 184 **W** tungsten 74 | 186 **Re** rhenium 75 | 190 **Os** osmium 76 | 192 **Ir** iridium 77 | 195 **Pt** platinum 78 | 197 **Au** gold 79 | 201 **Hg** mercury 80 | 204 **Tl** thallium 81 | 207 **Pb** lead 82 | 209 **Bi** bismuth 83 | [209] **Po** polonium 84 | [210] **At** astatine 85 | [222] **Rn** radon 86 |
| [223] **Fr** francium 87 | [226] **Ra** radium 88 | [227] **Ac*** actinium 89 | [261] **Rf** rutherfordium 104 | [262] **Db** dubnium 105 | [266] **Sg** seaborgium 88 | [264] **Bh** bohrium 107 | [277] **Hs** hassium 108 | [268] **Mt** meitnerium 109 | [271] **Ds** darmstadtium 110 | [272] **Rg** roentgenium 111 | | | | | | | |

# Fundamental Chemical Concepts

## Formulae

Chemical symbols are used with numbers to write **formulae** that represent the composition of compounds. Formulae are used to show:

- the different elements in a compound
- the number of atoms of each element in the compound
- the total number of atoms in the compound.

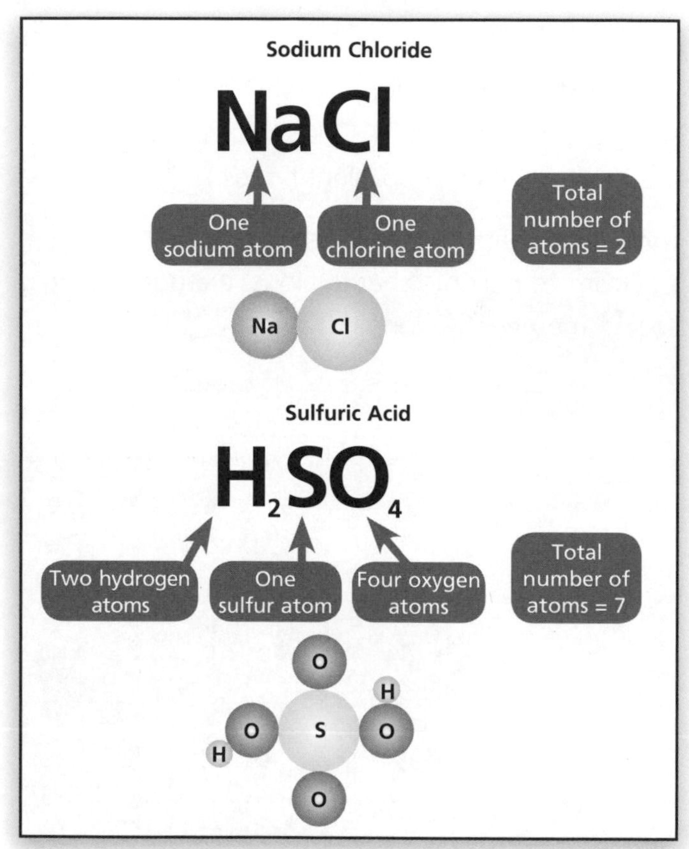


If there are brackets around part of the formula, everything inside the brackets is multiplied by the number outside the bracket.

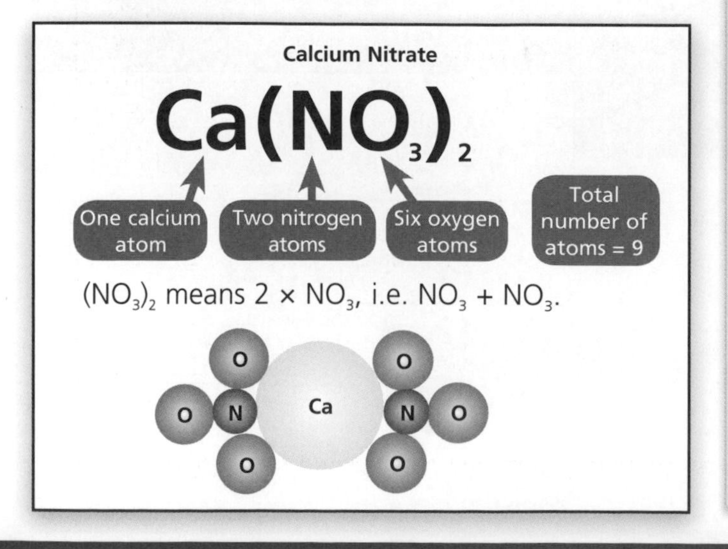


## Displayed Formulae

A **displayed formula** is another way to show the composition of a molecule.

A displayed formula shows:

- the different types of atom in the molecule, e.g. carbon, hydrogen
- the number of each different type of atom
- the covalent bonds between the atoms.

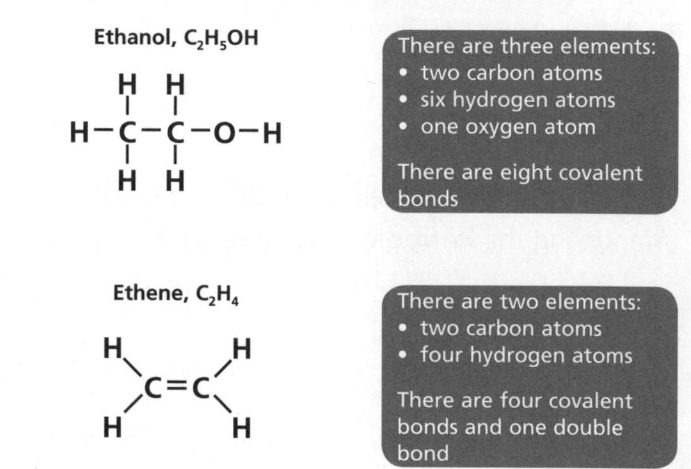


## Equations

In a chemical reaction, the substances that you start with are called **reactants**. During the reaction, the atoms in the reactants are rearranged to form new substances called **products**.

Chemists use equations to show what has happened during a chemical reaction. The reactants are on the left side of the equation, and the products are on the right.

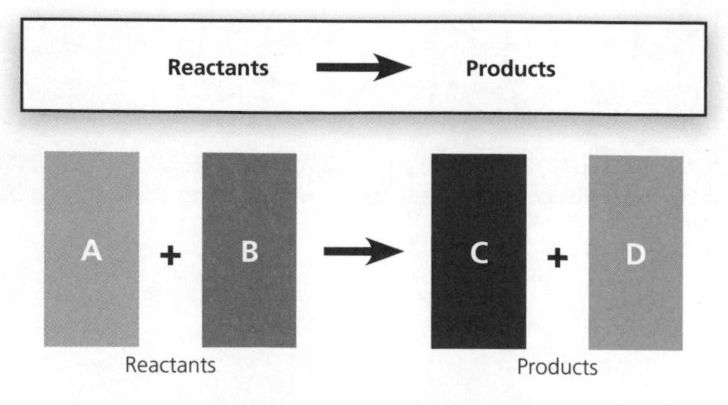


No atoms are lost or gained during a chemical reaction so equations must be **balanced**: there must always be the same number of atoms of each element on both sides of the equation.

## Writing Balanced Equations

**Example**

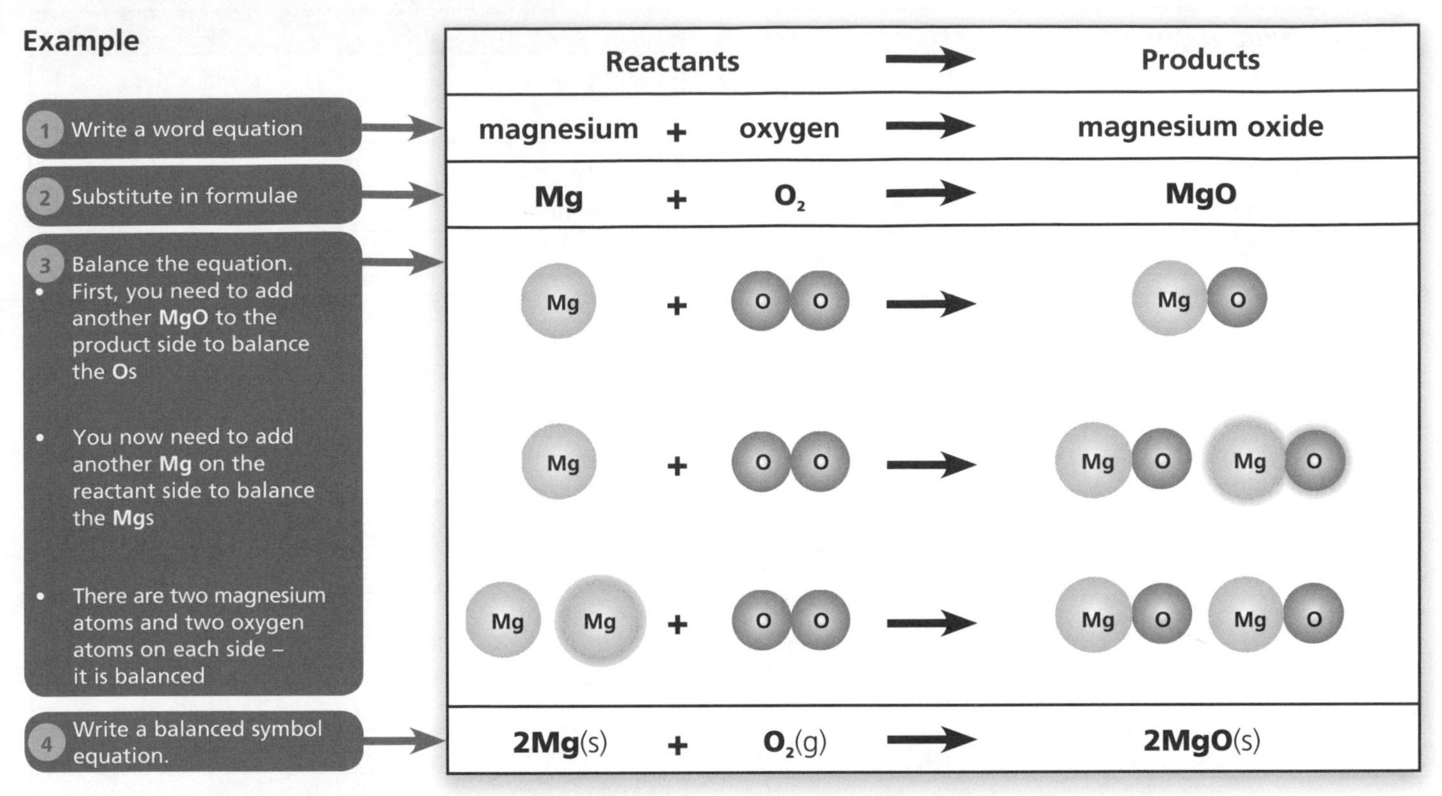


You may be asked to include the **state symbols** when writing an equation: (aq) for aqueous solutions, (g) for gases, (l) for liquids and (s) for solids.

HT You should be able to balance equations by looking at the formulae (i.e. without drawing the atoms).

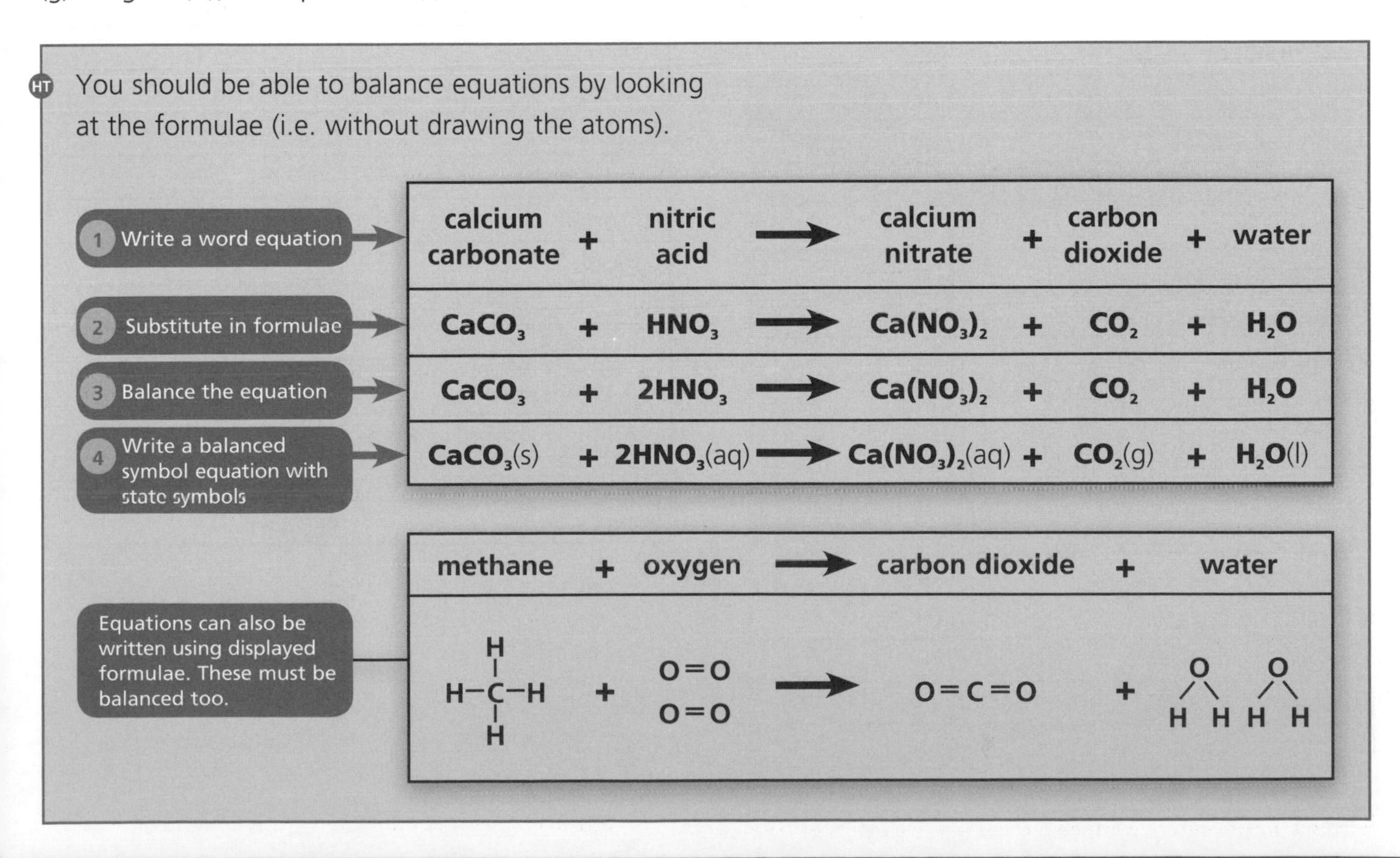

## Common Compounds and their Formulae

| Acids | |
|---|---|
| Ethanoic acid | $CH_3COOH$ |
| Hydrochloric acid | $HCl$ |
| HT Nitric acid | $HNO_3$ |
| HT Sulfuric acid | $H_2SO_4$ |
| **Carbonates** | |
| Calcium carbonate | $CaCO_3$ |
| HT Copper(II) carbonate | $CuCO_3$ |
| HT Iron(II) carbonate | $FeCO_3$ |
| HT Magnesium carbonate | $MgCO_3$ |
| HT Manganese carbonate | $MnCO_3$ |
| HT Sodium carbonate | $Na_2CO_3$ |
| HT Zinc carbonate | $ZnCO_3$ |
| **Chlorides** | |
| HT Ammonium chloride | $NH_4Cl$ |
| HT Barium chloride | $BaCl_2$ |
| HT Calcium chloride | $CaCl_2$ |
| HT Iron(II) chloride | $FeCl_2$ |
| HT Magnesium chloride | $MgCl_2$ |
| Potassium chloride | $KCl$ |
| HT Silver chloride | $AgCl$ |
| Sodium chloride | $NaCl$ |
| HT Tin(II) chloride | $SnCl_2$ |
| HT Zinc chloride | $ZnCl_2$ |
| **Oxides** | |
| Calcium oxide | $CaO$ |
| HT Copper(II) oxide | $CuO$ |
| HT Iron(II) oxide | $FeO$ |
| HT Magnesium oxide | $MgO$ |
| HT Manganese(II) oxide | $MnO$ |
| HT Sodium oxide | $Na_2O$ |
| HT Zinc oxide | $ZnO$ |
| **Hydroxides** | |
| HT Copper(II) hydroxide | $Cu(OH)_2$ |
| HT Iron(II) hydroxide | $Fe(OH)_2$ |
| HT Iron(III) hydroxide | $Fe(OH)_3$ |
| HT Lithium hydroxide | $LiOH$ |
| HT Potassium hydroxide | $KOH$ |
| HT Sodium hydroxide | $NaOH$ |

| Sulfates | |
|---|---|
| HT Ammonium sulfate | $(NH_4)_2SO_4$ |
| HT Barium sulfate | $BaSO_4$ |
| HT Calcium sulfate | $CaSO_4$ |
| HT Copper(II) sulfate | $CuSO_4$ |
| HT Iron(II) sulfate | $FeSO_4$ |
| HT Magnesium sulfate | $MgSO_4$ |
| HT Potassium sulfate | $K_2SO_4$ |
| HT Sodium sulfate | $Na_2SO_4$ |
| HT Tin(II) sulfate | $SnSO_4$ |
| HT Zinc sulfate | $ZnSO_4$ |
| **Others** | |
| Ammonia | $NH_3$ |
| Bromine | $Br_2$ |
| HT Calcium hydrogencarbonate | $Ca(HCO_3)_2$ |
| Carbon dioxide | $CO_2$ |
| Carbon monoxide | $CO$ |
| Chlorine | $Cl_2$ |
| HT Ethanol | $C_2H_5OH$ |
| HT Glucose | $C_6H_{12}O_6$ |
| Hydrogen | $H_2$ |
| Iodine | $I_2$ |
| HT Lead iodide | $PbI_2$ |
| HT Lead(II) nitrate | $Pb(NO_3)_2$ |
| HT Methane | $CH_4$ |
| Nitrogen | $N_2$ |
| Oxygen | $O_2$ |
| HT Potassium iodide | $KI$ |
| HT Potassium nitrate | $KNO_3$ |
| HT Silver nitrate | $AgNO_3$ |
| HT Sodium hydrogencarbonate | $NaHCO_3$ |
| HT Sulfur dioxide | $SO_2$ |
| Water | $H_2O$ |

## C3: Chemical Economics

This module looks at:

- How rates of reaction can be investigated.
- How temperature and concentration affect the rate of reaction.
- How surface area and catalysts affect the rate of reaction.
- Calculating relative formula mass, and calculating masses in reactions.
- Percentage yield and atom economy calculations.
- Exothermic and endothermic reactions; measuring reaction energy changes.
- Batch and continuous processes, manufacturing pharmaceuticals.
- Properties and uses of diamond, graphite, buckminster fullerene and nanotubes.

## Rates of Reaction

Chemical reactions occur at different rates: some are very slow, such as rusting, and some, for example burning and explosions, are very fast.

We can measure the time a reaction takes. Here are some times taken for a 2cm strip of magnesium to completely dissolve in different concentrations of an acid:

| Concentration | Reaction Time (s) |
|---|---|
| 0.01 | 205 |
| 0.02 | 114 |
| 0.03 | 72 |
| 0.04 | 33 |
| 0.05 | 17 |

You may be asked to plot results such as these on a graph where the axes have been prepared for you. You might be asked to interpret this information, for example, pick out the fastest reaction from this table or predict the time it would take if the acid concentration was 0.035 (i.e. about half way between 33 and 72, so 53 seconds).

The reaction stops because one of the **reactants** is all used up. In this case the magnesium is the **limiting reactant**, the one that is used up; the acid is in excess as some is left over at the end of the reaction. The amount of **product** is increased if more reactant is used; it is directly proportional to the amount of reactant, i.e. if you double the amount of reactant then the amount of product also doubles.

HT The amount of the limiting reactant you start with dictates the amount of product you will make. When there are more reactant particles, there is more reaction and more product particles are produced.

## Gas Reactions

When a gas is made you can follow the progress of the reaction by recording the mass lost as the gas escapes, or by collecting the gas and measuring the volume:

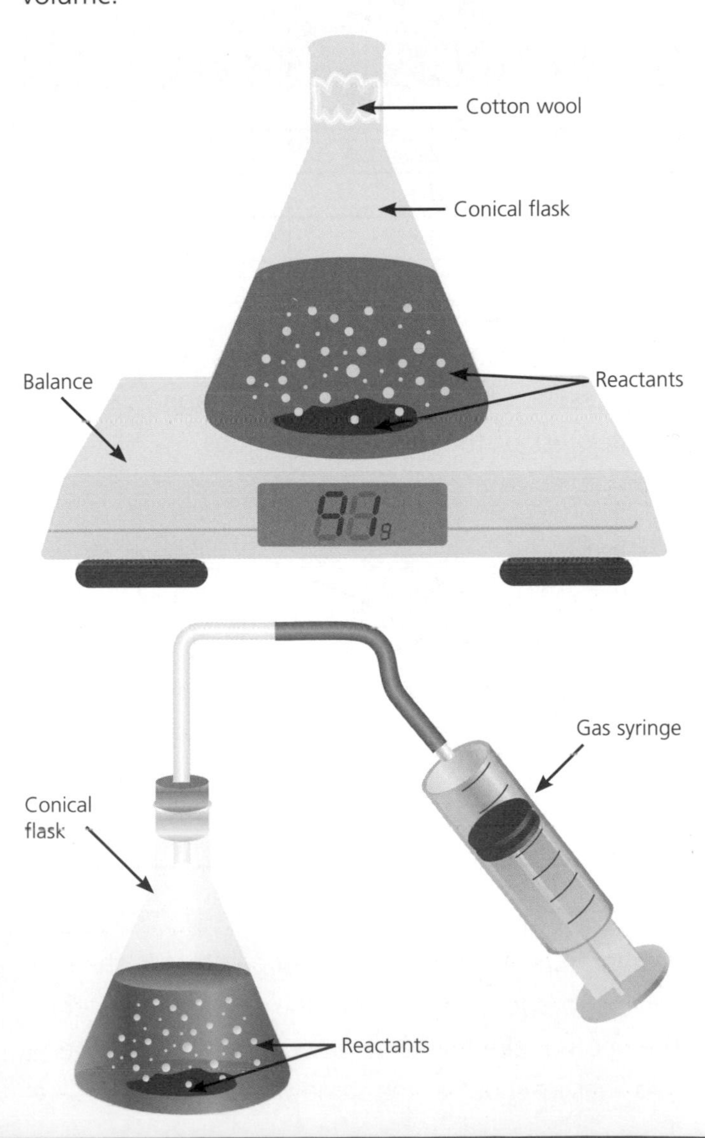

## Gas Reactions (cont)

Measuring the volume of gas made in a particular time would give you the rate of the reaction, or the fastest reaction would fill the syringe in the shortest time.

HT The rate of a reaction can be calculated to give the average volume of gas made in a particular time. The unit would be $cm^3/s$ or $cm^3/min$. Alternatively, if mass is used the units would be g/s or g/min.

## Analysing the Rate of Reaction

Two reactions were carried out at different temperatures and the following results were obtained:

| Time (s) | Volume of gas made at temperature A ($cm^3$) | Volume of gas made at temperature B ($cm^3$) |
|---|---|---|
| 0 | 0 | 0 |
| 30 | 22 | 18 |
| 60 | 40 | 35 |
| 90 | 54 | 49 |
| 150 | 62 | 57 |
| 180 | 62 | 62 |

These results can be sketched on a graph to show the progress of the chemical reactions.

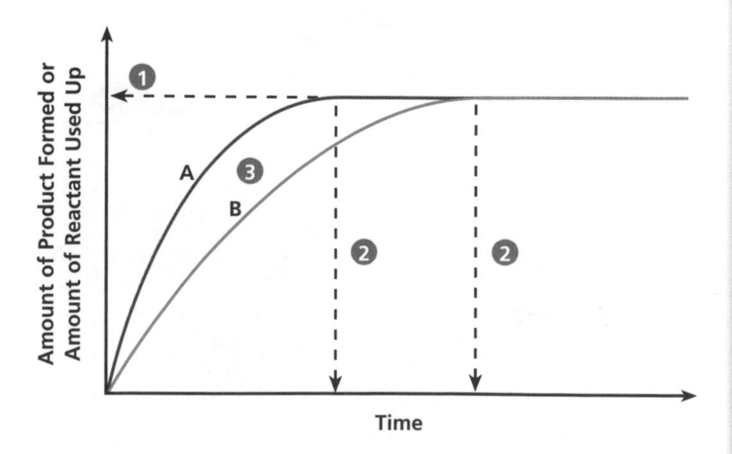


Temperature A is hotter than temperature B: the line for A is steeper showing that the reaction is faster. They both make the same amount of product; the two lines level out at the same value.

The graphs for concentration and pressure would be very similar with line A representing the higher concentration or the higher pressure.

You may be asked to plot results such as these on a graph where the axes have been prepared for you.

For example:

1. **Show how much product was made** by drawing a horizontal line from the highest point on the graph across to the $y$-axis.
2. **Show how long it takes to make the products**. The flat line on the graph indicates that the reaction is finished and that the products have been made. By drawing a vertical line down to the $x$-axis (time) from the flat line we can see how long this took.
3. **See which reaction is quicker** by comparing the steepness of the lines (the steeper the line, the quicker the reaction).

HT The graph below shows the progress of a chemical reaction. The points used to plot the graph are shown as crosses.

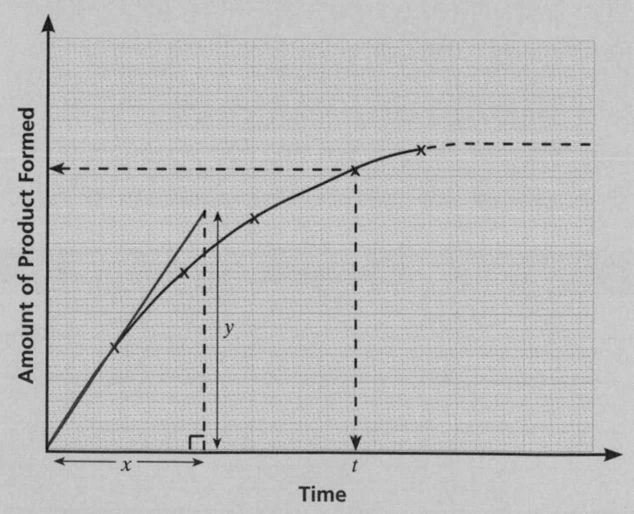


From the graph you can:

- calculate the initial rate of reaction by drawing a straight line following the start of the curve and working out $\frac{y}{x}$
- extend the curve by estimating the most likely path it will take next
- work out the amount of product formed by a time ($t$) for which we did not have a reading.

## Collisions and Rate of Reaction

The particles in a chemical reaction must collide together for a reaction to take place. The more collisions there are, the faster the reaction. The idea of collisions is used to explain how different factors affect the rate of a reaction.

## Temperature of the Reactants

| Lower temperature – lower rate | Higher temperature – higher rate |
|---|---|
| In a cold reaction mixture, the particles move quite slowly. The particles will collide with each other less often and with less energy, so fewer collisions will be successful. | If the temperature of the reaction mixture is increased, the particles will move faster. They will collide with each other more often and with greater energy, so many more collisions will be successful. |
| 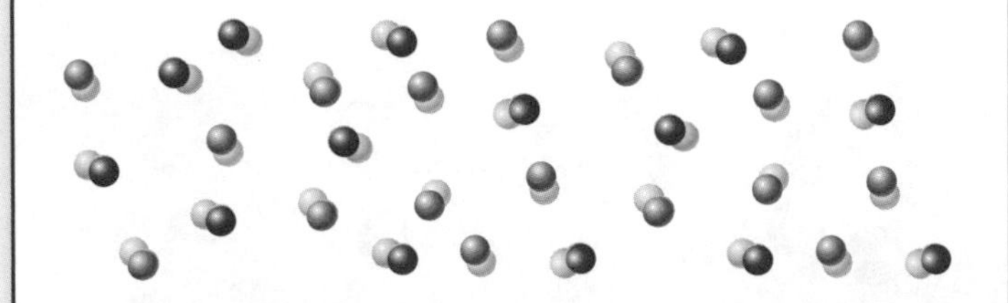 | 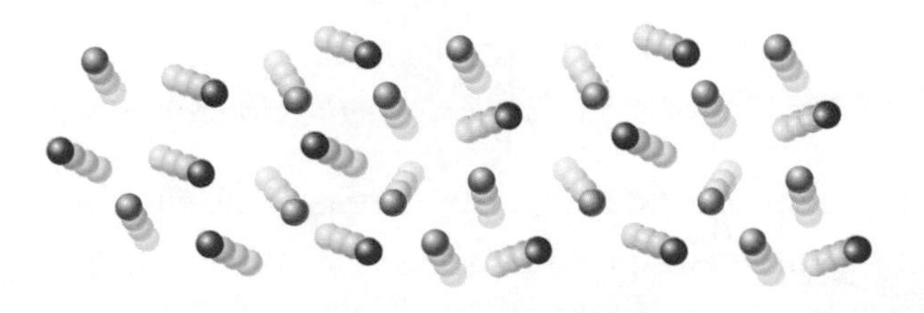 |

## Concentration of the Reactants

| Lower concentration – lower rate | Higher concentration – higher rate |
|---|---|
| In a reaction where one or both reactants are in low **concentrations**, the particles are spread out. The particles will collide with each other less often, resulting in fewer successful collisions. | Where there are high concentrations of one or both reactants, the particles are crowded close together. The particles will collide with each other more often, resulting in many more successful collisions. |
| 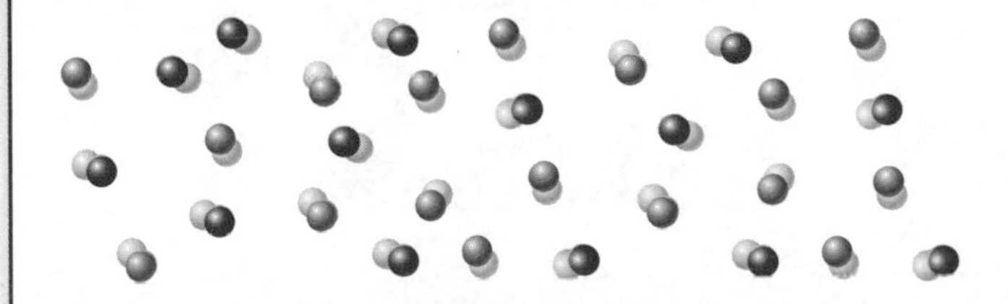 |  |

## Pressure of a Gas

| Lower pressure – lower rate | Higher pressure – higher rate |
|---|---|
| When a gas is under a low pressure, the particles are spread out. The particles will collide with each other less often resulting in fewer successful collisions. (This is like low concentration of liquid reactants.) | When the pressure is high, the particles are crowded more closely together. The particles collide more often, resulting in many more successful collisions. (This is like high concentration of liquid reactants.) |
| 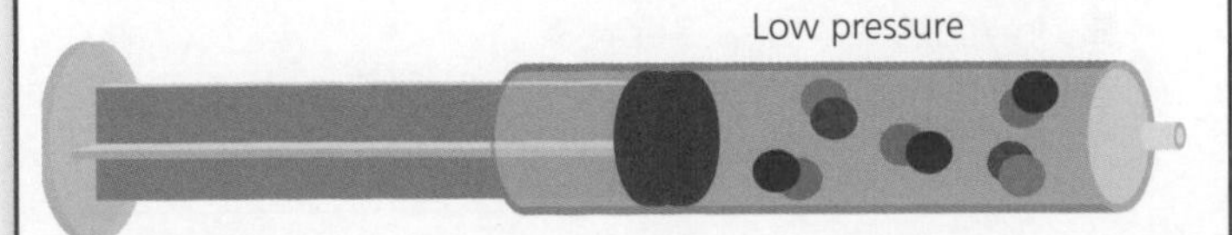  | 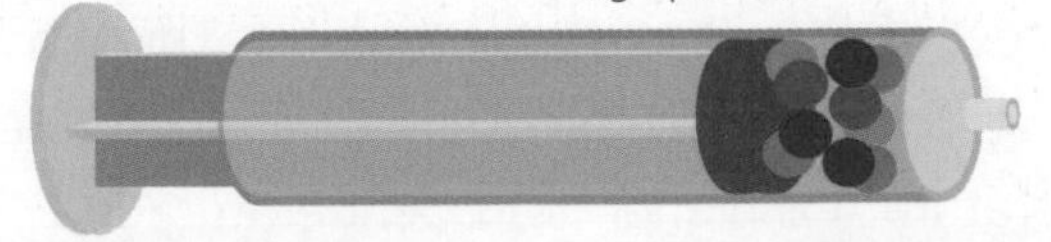  |

# Rate of Reaction (2)

## HT Collision Theory

Increasing **temperature** causes an increase in the kinetic energy of the particles, i.e. they move a lot faster. The faster the particles move, the greater the chance of them colliding, so the number of collisions per second increases. The more collisions there are between particles, the faster the reaction.

When the particles collide at an increased temperature they have more energy. When a collision has more energy, the chance of it causing a successful collision is increased (energetic collisions = more successful collisions).

Increasing **concentration** increases the number of particles in the same space, i.e. the particles are much more crowded together.

Increasing **pressure** in a gas reaction is very much like increasing the concentration; the particles are more crowded together and this increases the frequency of collisions between the particles.

The more crowded the particles are, the greater the chance of them colliding together, which increases the number of collisions per second. (*More frequent* collisions not just *more* collisions.)

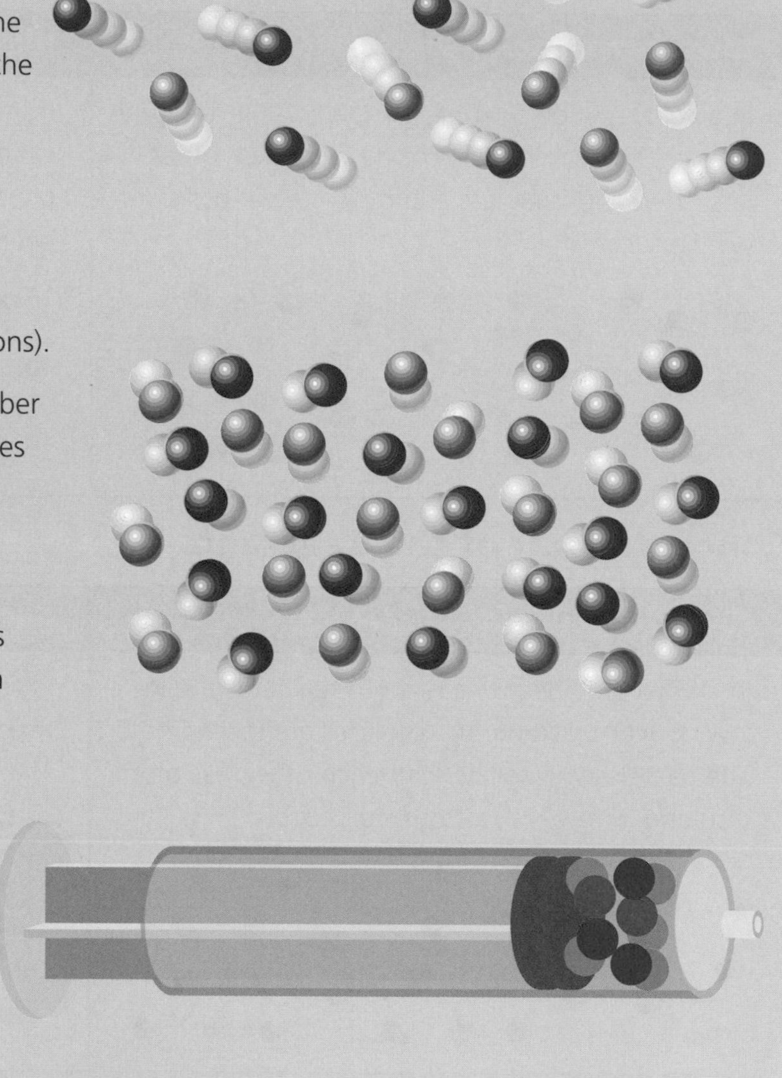

You may be asked to interpret information about the effect of changing temperature, concentration or pressure on a reaction. This could be in tables, graphs or a description. Use the guidelines on page 48 to help.

You may also be asked to sketch a graph showing the effect of changing these variables. This is a sketch graph showing the effect of temperature – temperature A is higher than temperature B:

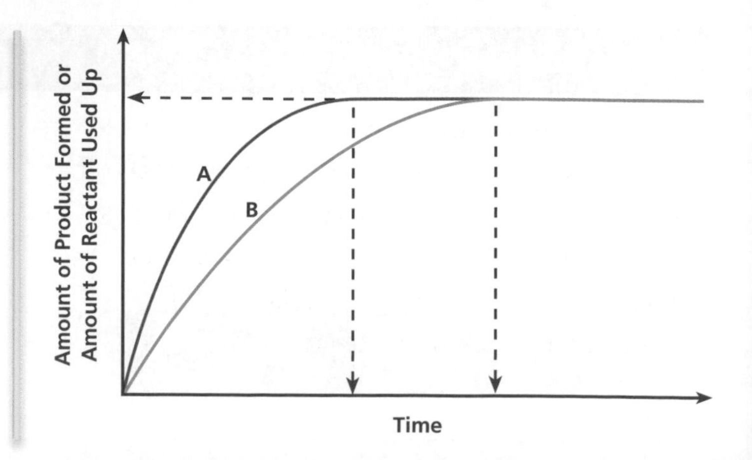

## Surface Area of Solid Reactants

Powdered solids react faster than lumps of the same reactant. A powdered reagent has a much larger surface area compared to lumps of the same material. There are more particles available on the surface for the other reactants to collide with. The greater the number of particles exposed, the greater the chance of them colliding together, which increases the reaction. (More collisions = faster reaction.)

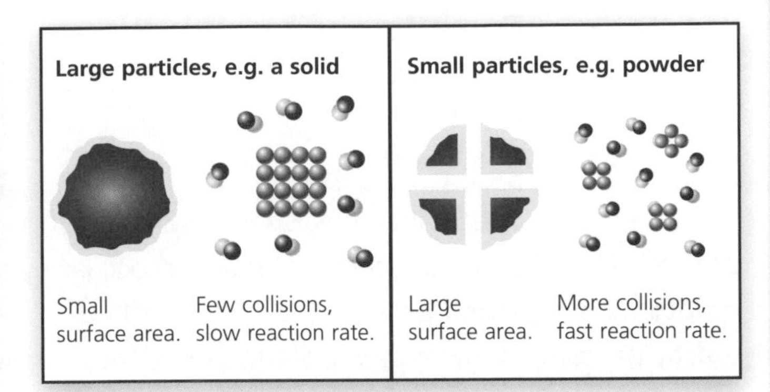


Gas and fine dust have the largest surface area of all and sometimes they react so fast it leads to an explosion, e.g. burning hydrogen or a custard powder explosion. An **explosion** is a very fast reaction where huge volumes of gas are made. Other materials that can explode are dynamite and TNT.

Factories that handle powders such as flour, custard powder or sulfur have to be very careful because the dust of these materials can mix with air and could cause an explosion if there is a spark. The factories have to prevent dust being produced and take precautions to ensure no spark is made that would ignite a dust–air mixture.

## Using a Catalyst

A **catalyst** is a substance that increases the rate of a reaction and is unchanged at the end of the reaction. Catalysts are very useful materials, as only a small amount of catalyst is needed to speed up the reaction of large amounts of reactant.

HT Catalysts (like reactants) are most effective when they have a large surface area. The greater the number of particles exposed, the greater the chance of them colliding together, which means the number of collisions per second increases. (*More frequent* collisions, not just *more* collisions.)

You may be asked to sketch a graph showing the effect of using a catalyst in an experiment, such as this:

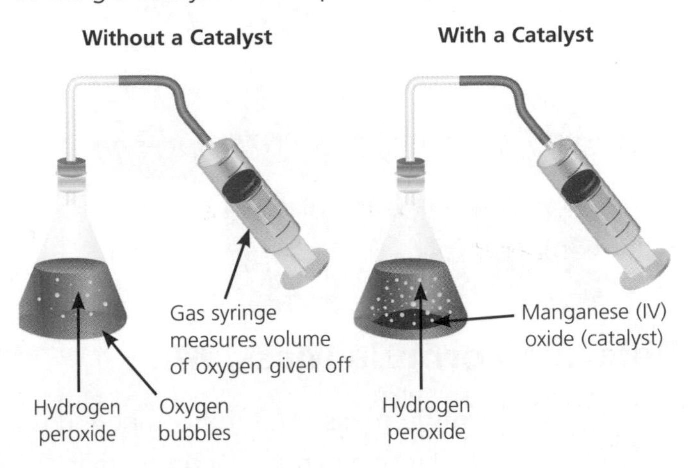


The sketch graph would look like this:

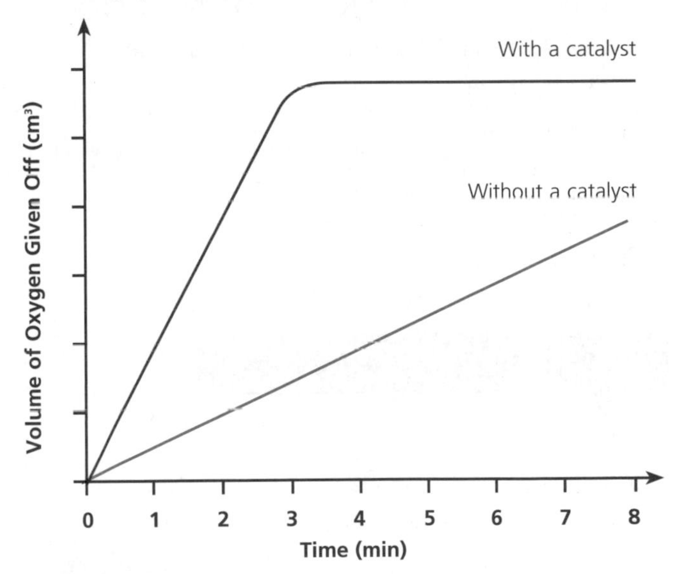


The sketch graph for changing the surface area would look very similar with the steeper line representing the larger surface area. You may be asked to interpret information about the effect of changing the surface area or using a catalyst on a reaction. This data could be in tables, graphs or a description.

# C3 Reacting Masses

## Relative Atomic Mass, $A_r$

**Atoms** are too small for their actual atomic mass to be of much use to us. A more useful measure is **relative atomic mass, $A_r$**. Use the Periodic Table to look up the relative atomic mass.

Each **element** in the Periodic Table has two numbers. The larger of the two numbers (top left below) is the $A_r$ of the element.

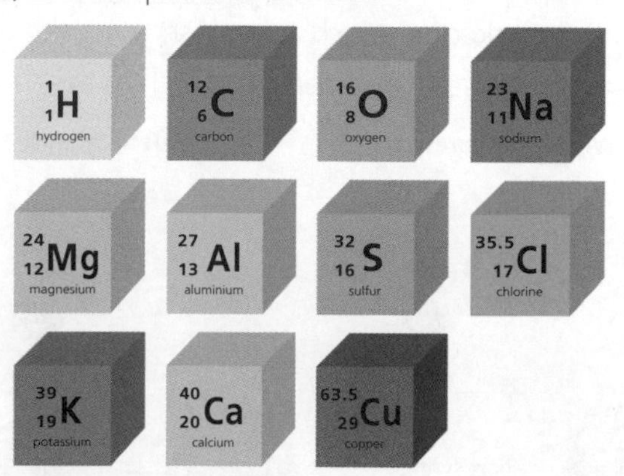


## Relative Formula Mass, $M_r$

The **relative formula mass, $M_r$**, of a compound is simply the $A_r$s of all its elements added together. To calculate the $M_r$, the formula of the compound and the $A_r$ of all the atoms involved are needed.

There are a number of ways to set out an $M_r$ calculation. Whichever method you choose, you should always show your working. The simplest method is shown below:

**Example 1**

Calculate the relative formula mass of $H_2SO_4$.

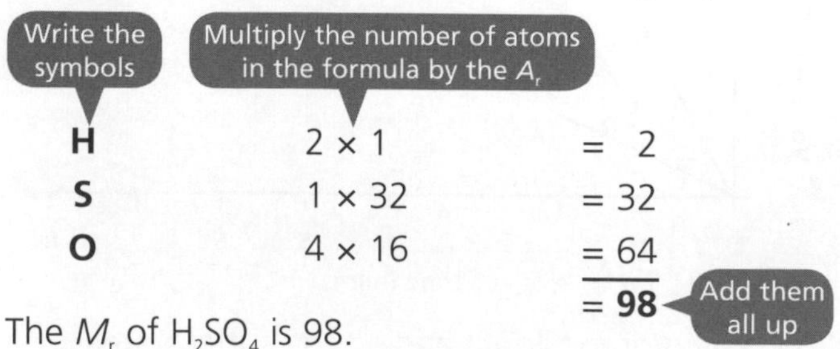


The $M_r$ of $H_2SO_4$ is 98.

**Example 2**

Calculate the relative formula mass of $Ca(OH)_2$.

| | | |
|---|---|---|
| **Ca** | 1 × 40 | = 40 |
| **O** | 2 × 16 | = 32 |
| **H** | 2 × 1 | = 2 |
| | | **= 74** |

## Conservation of Mass

The total mass of the starting materials (reactants) always equals the total mass of the substances produced (products). This is called the **principle of conservation of mass**. We can show this with a reaction such as when nitric acid ($M_r$ = 63) reacts with ammonia ($M_r$ = 17) to make ammonium nitrate ($M_r$ = 80):

$$HNO_3 + NH_3 \rightarrow NH_4NO_3$$

Total mass of reactants = 63 + 17 = 80
Total mass of product = 80

The total mass of reactants equals the total mass of products because when a chemical reaction occurs **no atoms are gained or lost**. You end up with exactly the same number as you started with; they are just rearranged into different substances.

Atoms are not created or destroyed during a chemical reaction. You can show that mass is conserved in any reaction, for example, burning propane:

$$C_3H_8 + 5O_2 \longrightarrow 3CO_2 + 4H_2O$$

Calculate the $M_r$ for each reactant and product:

$C_3H_8$ (3 × 12) + (8 × 1) = 44
$O_2$ (2 × 16) = 32

$CO_2$ 12 + (2 × 16) = 44
$H_2O$ (2 × 1) + 16 = 18

Multiply by the number of molecules of each and then add the masses together:

Reactants: 44 + (5 × 32) = 204

Products: (3 × 44) + (4 × 18) = 204

**Mass of reactants = Mass of products**

The idea that the total mass of reactants is the same as the total mass of products in a reaction can be used to calculate how much product is made or how much of a reactant is needed.

**Example 1**

When 20g of calcium carbonate is heated until it decomposes, 11.2g of calcium oxide is made. We can now work out the mass of carbon dioxide made:

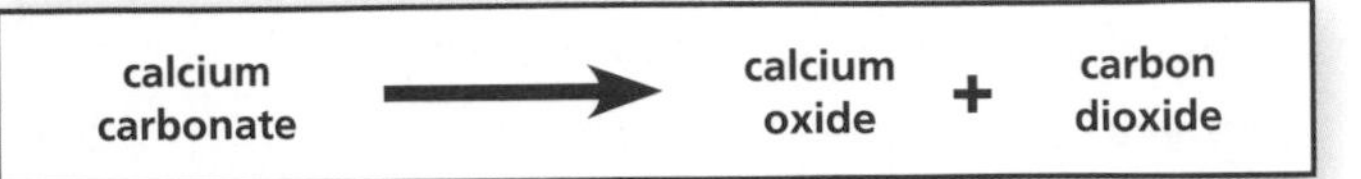


20g of reactant must make 20g of products
20g = 11.2g + mass of carbon dioxide

So the mass of carbon dioxide made
= 20g – 11.2g = **8.8g**

**Example 2**

What mass of oxygen is needed to react with 4.8g of magnesium to make 8g of magnesium oxide?

8g of reactants must make 8g of products
4.8g + mass of oxygen = 8g

So the mass of oxygen = 8g – 4.8g = **3.2g**

If you use more reactants then you will get more products. If you use twice as much reactant you get twice as much product. The mass of product obtained is directly proportional to the amount of reactant you start with. This idea can be used to calculate the mass of a reactant or product if you know the reacting ratios.

**Example 3**

From the previous example we know that 4.8g of magnesium makes 8g of magnesium oxide. We can double it:

2 × 4.8g = 9.6g of magnesium will make
2 × 8g = 16g of magnesium oxide

We can multiply by 10:
10 × 4.8g = 48g of magnesium will make
10 × 8g = 80g of magnesium oxide

**Example 4**

What mass of magnesium is needed to make 2g of magnesium oxide?

2g of magnesium is $\frac{2g}{8g} = \frac{1}{4}$ of the ratio.
$\frac{1}{4}$ of 4.8g = 1.2g, i.e. 1.2g of magnesium makes 2g of magnesium oxide.

## HT More Calculations

We sometimes need to be able to work out how much of a substance is used up or produced in a chemical reaction. To do this we need to know:

- the relative formula mass, $M_r$, of the reactants and products (or the relative atomic mass, $A_r$, of all the elements)
- the balanced symbol equation for the reaction.

By substituting the $M_r$s into the balanced equation, we can work out the ratio of mass of reactant to mass of product and apply this to the question.

**Example 1**

When calcium carbonate ($CaCO_3$) is heated, it produces calcium oxide (CaO) and carbon dioxide ($CO_2$). How much calcium oxide can be produced from 50kg of calcium carbonate?

(Relative atomic masses: Ca = 40, C = 12, O = 16.)

Write down the equation.

$$CaCO_3(s) \xrightarrow{heat} CaO(s) + CO_2(g)$$

Work out the $M_r$ of each substance.

$$40 + 12 + (3 \times 16) \rightarrow (40 + 16) + [12 + (2 \times 16)]$$

Check that the total mass of reactants equals the total mass of the products. If they are not the same, check your work.

$$100 \rightarrow 56 + 44 \checkmark$$

Since the question only mentions calcium oxide and calcium carbonate, you can now ignore the carbon dioxide. This gives the ratio of mass of reactants to mass of products.

100 : 56

Apply this ratio to the question:

If 100kg of $CaCO_3$ produces 56kg of CaO,
then 1kg of $CaCO_3$ produces $\frac{56}{100}$ kg of CaO
and 50kg of $CaCO_3$ produces $\frac{56}{100} \times 50 =$ **28kg** of CaO.

# C3 Percentage Yield and Atom Economy

## Percentage Yield

Whenever a reaction takes place, the starting materials (i.e. the reactants) produce new substances (i.e. the products). The greater the amount of reactants used, the greater the amount of products formed.

**Percentage yield** is a way of comparing the actual amount of product made (the actual yield) to the amount of product theoretically expected, which is the predicted yield. It is calculated using the following formula:

$$\text{Percentage yield} = \frac{\text{Actual yield}}{\text{Predicted yield}} \times 100$$

- A 100% yield means that no product has been lost, i.e. the actual yield is the same as the predicted yield.
- A 0% yield means that no product has been made, i.e. the actual yield is zero.

**Example 1**

A reaction was carried out to produce the **salt** magnesium sulfate. The correct amounts of reactants were added and allowed to react. The resulting solution was evaporated and the salt crystals were obtained by filtration. The predicted yield of magnesium sulfate was 7g but the actual yield was 4.9g.

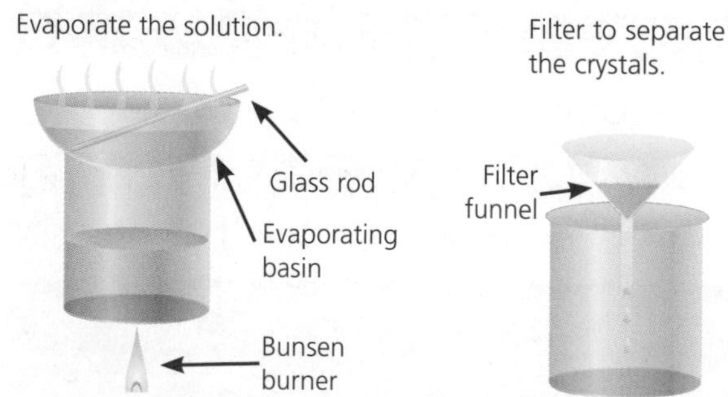


Calculate the percentage yield of magnesium sulfate.

$$\text{Percentage yield} = \frac{\text{Actual yield}}{\text{Predicted yield}} \times 100$$

$$= \frac{4.9\text{g}}{7\text{g}} \times 100 = \mathbf{70\%}$$

The percentage yield of magnesium sulfate was 70% which means that some magnesium sulfate was lost during the process. It could have been lost during evaporation, filtration, the transfer of liquids and / or heating.

## Atom Economy

Calculating the **atom economy** is an alternative method of deciding how effective a reaction is. The atom economy is a way of measuring the amount of atoms that are wasted when a chemical is manufactured. It is calculated using the following formula:

$$\text{Atom economy} = \frac{M_r \text{ of the desired products}}{\text{Total } M_r \text{ of all products}} \times 100$$

A 100% atom economy means that all the reactant atoms have been converted into desired product.

The higher the atom economy, the 'greener' the process (less waste).

**Example 1**

Limestone ($CaCO_3$, $M_r = 100$) is heated to make the useful product calcium oxide ($CaO$, $M_r = 56$). The other product is carbon dioxide ($CO_2$, $M_r = 44$). What is the atom economy for this important industrial process?

$$\text{Atom economy} = \frac{M_r \text{ of the desired products}}{\text{Total } M_r \text{ of all products}} \times 100$$

$$= \frac{56}{56 + 44} \times 100 = \mathbf{56\%}$$

You may be asked to make judgements about reactions such as this; with a 56% atom economy this is quite a wasteful process and not very green. A typical yield for the process is 95% and you would then decide that the actual reaction is quite effective.

HT An industrial process needs as high a percentage yield as possible in order to:
- reduce costs
- minimise wasting reactants.

The atom economy also needs to be as high as possible in order to:
- reduce unwanted products
- make the process more sustainable.

## Exothermic and Endothermic Reactions

Many reactions are accompanied by a **temperature rise**. These are known as **exothermic** reactions because heat energy is **given out** to the surroundings.

Some reactions are accompanied by a **fall in temperature**. These reactions are known as **endothermic** reactions because heat energy is **taken in** from the surroundings (the reaction absorbs energy).

## Comparing Fuels

The equipment shown in the diagram below can be used to compare the amounts of heat energy released by the combustion of different fuels.

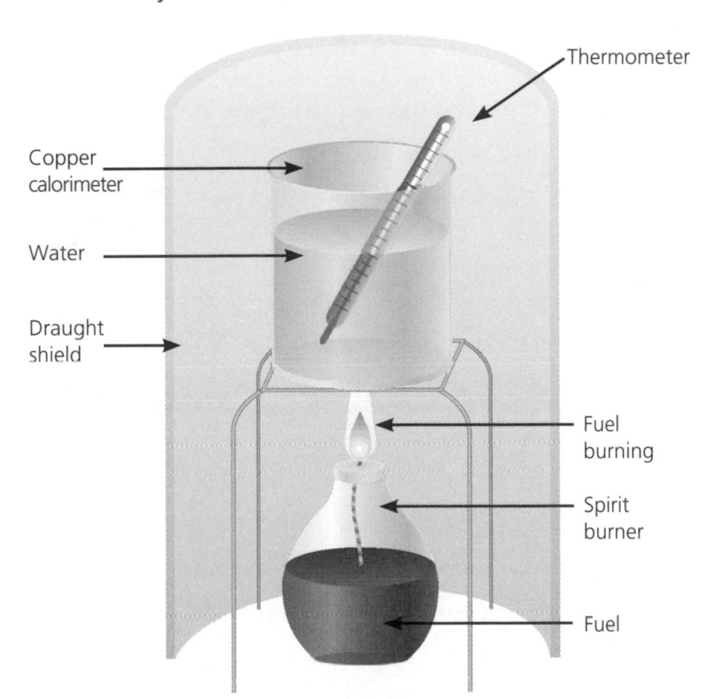


The greater the rise in the temperature of the water, the greater the amount of energy, in joules (J) or kilojoules (kJ), released from the fuel being used.

To make meaningful comparisons we would need to carry out a **fair test** each time. We would need to:
- use the same mass (volume) of water
- use the same calorimeter
- have the burner and calorimeter the same distance apart
- burn the same mass of fuel.

The formula used to work out the change in temperature (°C) is:

**Temperature change** (°C) = **Final temperature of water** (°C) – **Start temperature of water** (°C)

If you burn the same mass of each fuel, the fuel that produces the largest temperature rise releases the most energy.

1g of three different fuels were burned in the above apparatus. Here are the results:

| Fuel | Start Temp. °C | Final Temp. °C | Temperature change °C |
|---|---|---|---|
| Propanol | 19 | 32 | 13 |
| Ethanol | 19 | 30 | 11 |
| Pentane | 18 | 34 | |

You may be asked to calculate the missing temperature change (i.e. 34°C – 18°C = 16°C) and which fuel gave out the most energy (i.e. pentane because it had the largest temperature change).

## Breaking and Making Bonds

In a chemical reaction:
- breaking bonds is an **endothermic** process
- making bonds is an **exothermic** process.

HT

Chemical reactions that require more energy to break bonds than is released when new bonds are made are **endothermic** reactions.

Chemical reactions that release more energy when bonds are made than is required to break bonds are **exothermic** reactions.

For example, bottled gas (propane) burns in air to release lots of energy. The displayed formula equation for the burning of propane is shown on the next page.

## HT Breaking and Making Bonds (cont)

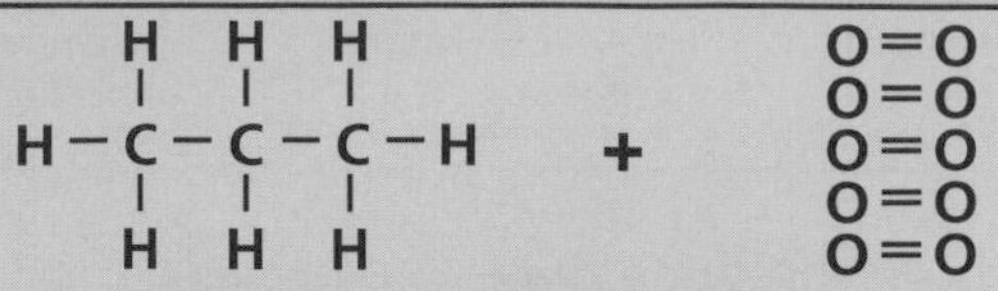

All the bonds in propane and oxygen have to be broken. Energy needs to be taken in to break the bonds.

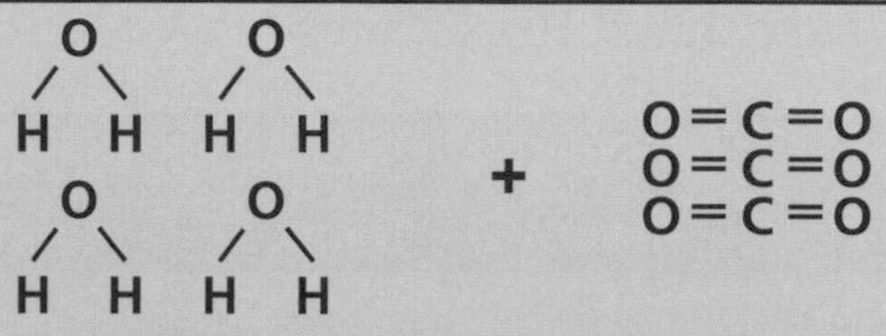

All the bonds in water and carbon dioxide have to be made. Energy is given out when bonds are made.

When propane burns, more energy is given out when the new bonds are made than is taken in to break the old bonds at the start. The **overall energy change is exothermic**.

## Calculating Energy Changes

In order to compare fuels, we need to work out the amount of energy transferred by the fuel to the water, and the amount of energy transferred per gram of fuel burned. The results from an experiment with hexane are as follows:

| | Start | End |
|---|---|---|
| Mass of burner and hexane | 187.60g | 187.34g |
| Temperature of water | 22°C | 34°C |

Mass of hexane burned (187.60 – 187.34) = **0.26g**

Rise in temperature of water (34°C – 22°C) = **12°C**

Mass of water in calorimeter = **200g**

The amount of energy transferred to the water can be calculated using the following formula:

**Energy supplied to warm the water = Mass of water × Specific heat capacity × Temperature change**

Energy transferred = $m \times c \times \Delta T$

*N.B. You do not need to remember this formula.*

Energy supplied = 200g × 4.2J/g/°C × 12°C
= **10 080 joules**

*N.B. Specific heat capacity is a constant that is specific to a particular liquid. For water it has a value of 4.2J/g/°C.*

HT The energy transferred per gram of fuel burned can be calculated using the following formula:

$$\textbf{Energy per gram} = \frac{\textbf{Energy supplied}}{\textbf{Mass of fuel burned}}$$

*N.B. You need to remember this formula.*

$$\text{Energy per gram} = \frac{10\,080}{0.26\text{g}}$$
= **38 769J/g**

You may be asked to use the energy transferred equation to calculate the temperature change. A fuel transferred 12 600J of energy when heating 200g of water. Calculate the temperature change.

Energy transferred = $m \times c \times \Delta T$ and so
$12\,600 = 200 \times 4.2 \times \Delta T$

Rearrange the equation.

$$\Delta T = \frac{12\,600}{200 \times 4.2} = 15°\text{C}$$

The equation can also be used to calculate the mass of the water heated up. The temperature of a beaker of water went up by 18°C when 11 340J of energy was used. Calculate the mass of the water.

Energy transferred = $m \times c \times \Delta T$ and so
$11\,340 = m \times 4.2 \times 18$

Rearrange the equation.

$$m = \frac{11\,340}{4.2 \times 18} = 150\text{g}$$

## Batch and Continuous

In a **batch process** the reactants are put into a reactor, the reaction happens and then the product is removed. Medicines and pharmaceutical drugs are often made in batches. Batch processes make a product on demand, make a product on a small scale, can be used to make a variety of products and are more labour intensive because the reactor needs to be filled, emptied and cleaned.

In a **continuous process**, e.g. making ammonia, reactants are continually being fed into a large reactor and the product is continually being produced at the same time (like a conveyer belt). Continuous processes operate all the time and run automatically, make a product on a large scale and are dedicated to just one product.

## Making Medicines

The materials used to make a medicine can be **manufactured** (synthetic) or they can be extracted from **natural sources** such as plants. The steps needed to extract a small amount of material from a plant source are as follows:

1. **Crushing** – the plant material is crushed using a mortar and pestle.
2. **Dissolving** – a suitable solvent is added to dissolve the material. This could be **boiled** to improve the extraction.
3. **Chromatography** – a concentrated solution of the material is spotted onto chromatography paper and allowed to separate.

## Developing Medicines

It takes a long time – over 10 years – from discovering a material that will act as a medicine, to being able to use it on patients. **Research** needs to be carried out into new pharmaceutical materials. This cannot be automated (carried out by machines) as decisions need to be made, so highly qualified scientists are needed. This means that labour costs are high. Further research is then carried out to **develop** the drug to increase its effectiveness before it is **tested** to ensure it works properly, is safe to use and has no serious side-effects. The medicine must then be approved for use and must satisfy all the **legal requirements** before it can be sold.

Medicines are expensive because the materials could be rare or may require complex methods to extract the **raw materials** (starting materials) from plants. Medicines are made in small quantities and it is not possible to totally automate the manufacturing process. Medicines must be as pure as possible to avoid side effects caused by impurities. Making medicines is labour intensive and, therefore, staff costs are high. The marketing of a new medicine is also very expensive.

HT The research into, and development of, a new pharmaceutical material takes a few years. Hundreds of similar molecules have to be made and tested to find the one that works best. Sometimes the best molecules have too many or serious side effects, so more development is needed. Once a material is developed as a medicine it has to undergo lots of testing to ensure it is better than other available medicines and is safe to use. There are many regulations on how it can be tested, such as when, how and if it can be tested on animals. Ultimately it has to be tested on human volunteers. Some countries have very strict legal rules that a new medicine must satisfy before it can be put on the market.

The table is a summary of how medicines can be tested to make sure they are pure:

| Test | Result if Pure | Result if Impure |
|---|---|---|
| Melting point | Single, sharp temperature. | Melts over range of temperatures. |
| Boiling point | Single, sharp temperature. | Boils over range of temperatures. |
| Chromatography | Single dot on paper. | Two or more dots on paper. |

You may be given some information about a medicine and be asked to decide if it is pure using the above test results.

## Carbon

There are three main **allotropes** of carbon:

- Diamond (see below)
- Graphite (see below)
- Buckminster fullerene (buckyball – see page 59).

Allotropes are different forms of the same element in the same physical state. The atoms of the element are arranged in a different molecular structure. The allotropes of carbon are all solids.

### Diamond

Diamond has a rigid structure.

- It does not conduct electricity.
- It is **insoluble** in water.
- It is used in jewellery because it is colourless, clear (transparent) and lustrous (shiny).
- It can be used in cutting tools because it is very hard and has a very high melting point.

**The Structure of Diamond**

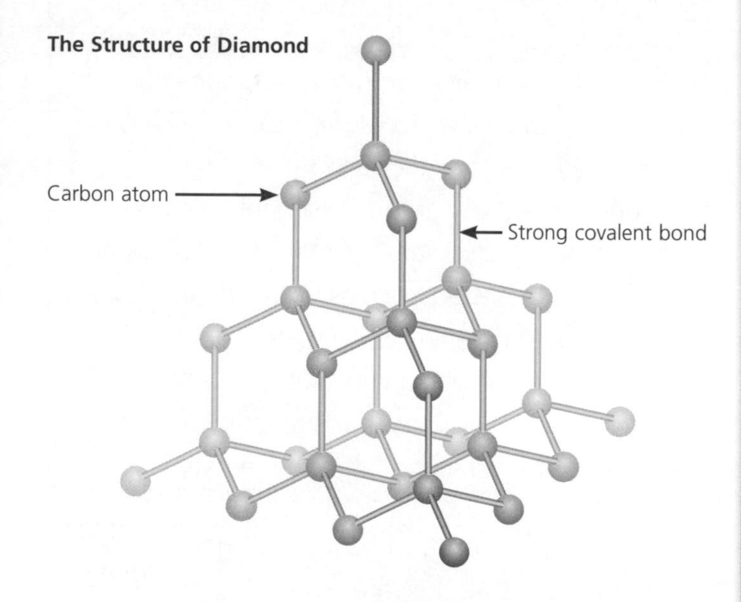


### Graphite

Graphite has a layered structure.

- It is insoluble in water.
- It is black, which is why it is used in pencils.
- It is lustrous and opaque (light cannot travel through it).
- It conducts electricity and has a very high melting point, so is used to make electrodes for electrolysis.
- It is slippery, so it is used in lubricants.

**The Structure of Graphite**

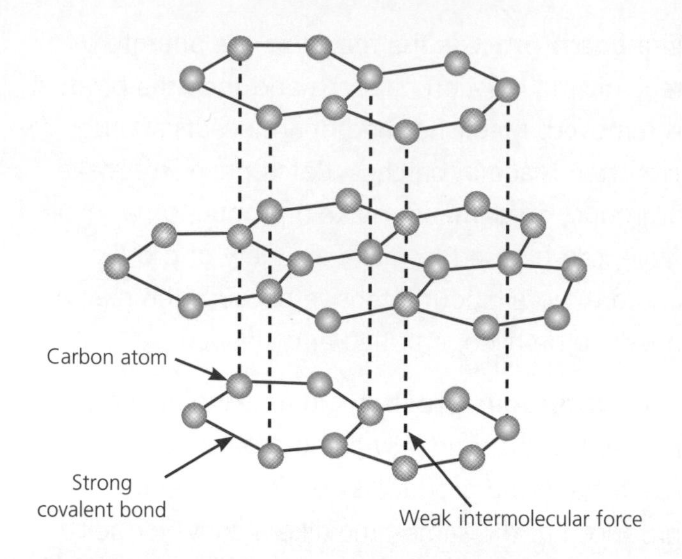


HT

### More on Diamond

Diamond is a giant molecule made of carbon atoms that are bonded to four other carbon atoms by strong covalent bonds. The giant structure has a large number of covalent bonds and this results in diamond having a high melting point because it needs lots of energy to break the bonds. It does not have any free electrons so it does not conduct electricity. Unlike graphite, it does not have separate layers and, because there are strong covalent bonds between the carbon atoms, it is very hard.

### More on Graphite

Graphite is a giant molecule that exists in layers of carbon atoms bonded to three other carbon atoms by strong covalent bonds. The layers are held together by weak intermolecular forces, allowing each layer to slide easily, so graphite can be used as a lubricant. The presence of free (delocalised) electrons in graphite results in it being an electrical conductor. It also has a high melting point because the giant structure has many strong covalent bonds to break and lots of energy is needed to do this.

Diamond and graphite are giant molecular structures – many trillions of atoms joined together in a network by covalent bonds.

## Nanochemistry

**Nanochemistry** deals with materials on an atomic scale (i.e. individual atoms), whereas chemistry is usually concerned with much larger quantities. Two forms of carbon that nanochemists are interested in are buckminster fullerene and nanotubes.

### Buckminster Fullerene

Buckminster fullerene ($C_{60}$) consists of 60 carbon atoms arranged in a sphere.

**Structure of Buckminster Fullerene**

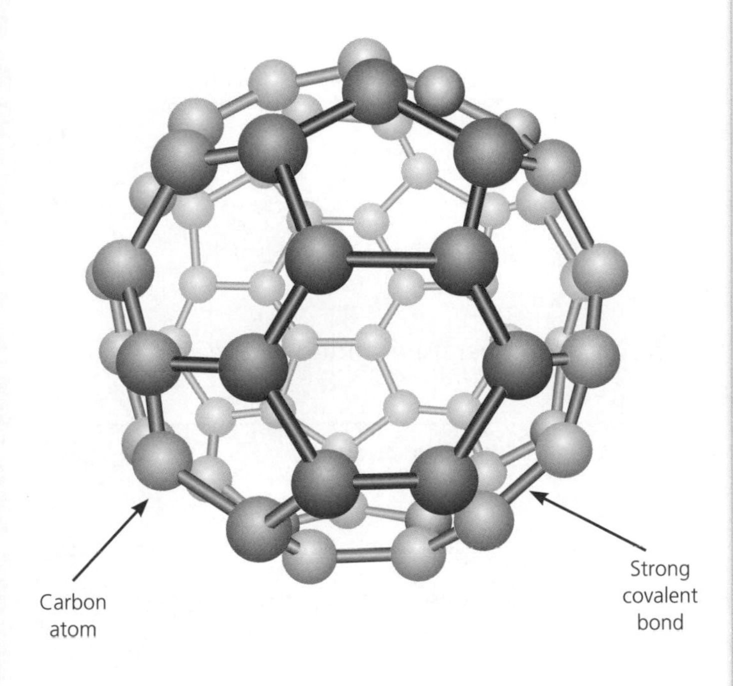


### Nanotubes

The discovery of buckminster fullerene led chemists to investigate building similar structures.

In the early 1990s the first **nanotubes** were made by joining fullerenes together. They look like sheets of graphite hexagons curled over into a tube.

Nanotubes conduct electricity and are very strong. They are used to:

- reinforce graphite tennis rackets because of their strength
- make connectors and semiconductors in the most modern molecular computers because of their electrical properties
- develop new, more efficient industrial catalysts.

**Structure of a Nanotube**

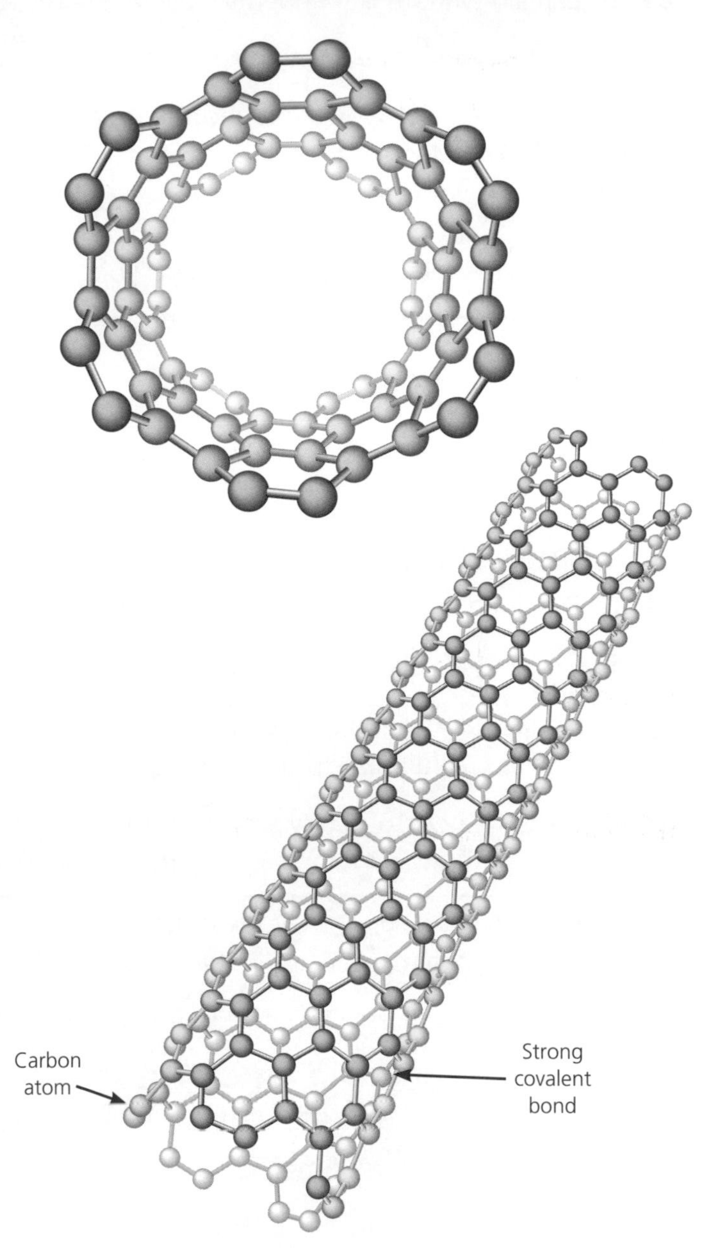


Fullerenes and nanotubes can be used to cage other molecules because they are the perfect shape to trap other substances inside them. The caged substances can be:

- **drugs**, e.g. a major new HIV treatment uses buckyballs to deliver a material which disrupts the way in which the HIV virus works

HT

- **catalysts** – by attaching catalyst material to a nanotube, a massive surface area can be achieved, making the catalyst very efficient.

# Exam Practice Questions

1 **a)** Explain what must happen to reactant particles for a reaction to take place. **[2]**

**b)** Explain why some collisions do not result in a successful reaction. **[2]**

2 Describe the precautions that have to be taken in flour mills to prevent explosions.
Use your knowledge of rates of reaction to explain the danger of explosion. **[6]**

*The quality of written communication will be assessed in your answer to this question.*

3 Calculate the relative formula mass ($M_r$) of:

**a)** sodium carbonate, $Na_2CO_3$ **[1]**

**b)** ethanol, $C_2H_5OH$ **[1]**

**c)** In the following reaction, 12g of carbon reacts with 32g of oxygen: **carbon + oxygen ➔ carbon dioxide**

**i)** Calculate the mass of carbon dioxide made in the reaction. **[1]**
**ii)** What mass of oxygen is needed to burn 1g of carbon? Give your answer to 2 significant figures. **[2]**

4 **a)** Diamond and graphite are both forms of which element? **[1]**

**b)** Explain why diamond is used in cutting tools. **[2]**

5 A teacher was making ammonium sulfate in a school laboratory. Decide whether this is a batch process or a continuous process and give a reason why. **[2]**

6 **a)** David heated 100g of water by burning 5g of petrol. The water temperature started at 21°C and went up to 47°C. Calculate the temperature change. **[1]**

HT **b)** Use David's results to calculate how much energy was made by burning the petrol. **[2]**
(Specific heat capacity of water = 4.2 J/g/°C)

7 Use ideas about structure and bonding to explain why graphite can conduct electricity. **[3]**

8 Magnesium reacts with hydrochloric acid to make hydrogen and magnesium chloride according to the equation: $\mathbf{Mg + 2HCl \rightarrow H_2 + MgCl_2}$

Calculate the mass of magnesium chloride made by reacting 9.6g of magnesium. **[2]**

9 Look at the graph below. The curve from the original reaction is labelled.

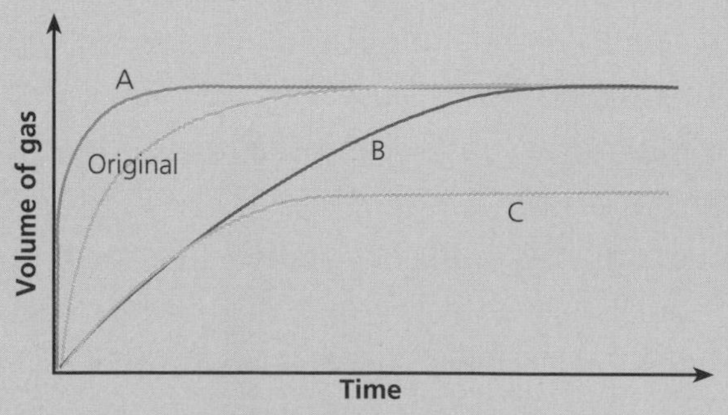

Select the curve on this graph that represents the curve showing the same reaction but using:

**a)** a lower original temperature. **[1]**
**b)** half the original amount of reactants. **[1]**
**c)** Describe how you would use the graph to compare the speed of the original reaction with the speed of reaction B. **[2]**

**C4: The Periodic Table**

This module looks at:
- The structure of the atom, protons, neutrons and electrons; isotopes.
- How metals and non-metals bond together, and the properties of ionic substances.
- Groups and periods, and properties of molecules.
- Properties of Group 1 metals, their reaction with water and flame tests.
- Properties of Group 7 non-metals, including their reaction with alkali metals.
- Properties of transition metals, decomposition of carbonates, and precipitation reactions.
- Metallic structure, properties of metals and superconductors.
- Sources and treatment of drinking water, pollution and testing for dissolved ions.

## The Discovery of Atoms

The model of atomic structure has changed over time as new evidence has been discovered. The idea that everything is made from atoms was first put forward in ancient Greece but it was not until **John Dalton** published his 'Atomic Theory' in 1803 that the idea started to gain acceptance. He imagined atoms to look like billiard balls. His main ideas were:
- Elements are made from very tiny particles called atoms, which cannot be made or broken up.
- Atoms of the same element are identical and different to other elements' atoms.
- Atoms combine to make compounds.

We now know that the theory was not totally correct. **J. J. Thomson** discovered that even smaller particles could be pulled out of atoms using a very high voltage (the discovery of the electron). He pictured the atom to be solid like a currant bun and his experiments removed the currants! A few years later, **Ernest Rutherford** proved that the atom was not solid but had a very dense centre called a nucleus. **Niels Bohr** combined his work on the orbits of electrons to develop the idea of the solar system atom, the nucleus in the centre like the Sun with the electrons orbiting around it like planets.

The ideas at each stage did not fully explain the behaviour of atoms. All the evidence had to be tested and predictions had to be made that would all have to be correct before a new theory was accepted.

HT Each development was a step closer to understanding the atom and helped other scientists with their work. Not all experiments went to plan: Rutherford's assistants, Geiger and Marsden, expected the $\alpha$ particles they were firing at gold foil to punch through in a straight line. A few were deflected and some even bounced straight back leading to the idea that the $\alpha$ particles were passing very close to, or hitting, the nucleus of the gold atoms.

## Structure of an Atom

All substances are made up of very small particles called **atoms** with a very low mass. Atoms have a central **nucleus**. The nucleus is positively charged because it is made up of **protons** which are positively charged and **neutrons** which are neutral. The space around the nucleus is occupied by negatively charged **electrons** arranged in shells. Overall, an atom has **no charge**.

HT An atom has the same number of protons and electrons, so they cancel out each other's charges. The mass of an atom is about $10^{-23}$g and it has a radius of $10^{-10}$m.

**A Fluorine Atom**

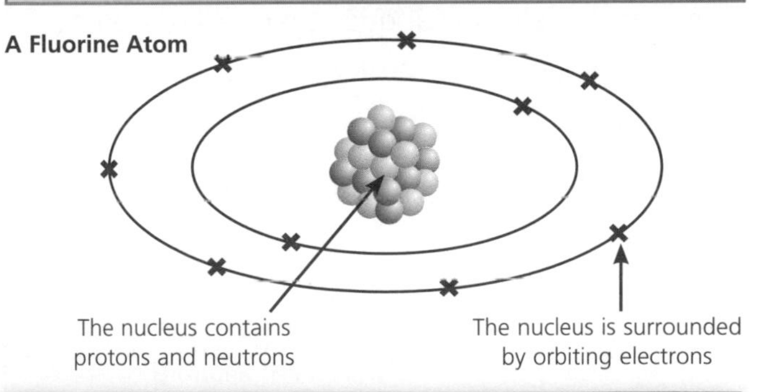


| Atomic Particle | Relative Charge | Relative Mass |
|---|---|---|
| Proton | +1 | 1 |
| Neutron | 0 | 1 |
| Electron | –1 | 0.0005 (zero) |

# C4 Atomic Structure

## Elements and Compounds

An **element** is a substance that cannot be broken down chemically and contains only one type of atom.

A **compound** is a substance that contains at least two elements that are **chemically combined**. You can find out which elements make up a compound by looking at the compound's **formula** and identifying the elements from the Periodic Table, for example:

- sodium chloride (NaCl) contains the elements sodium (Na) and chlorine (Cl)
- potassium nitrate ($KNO_3$) contains the elements potassium (K), nitrogen (N) and oxygen (O).

## Atomic Number

The numbers next to the elements in the Periodic Table give information about the element.

The elements in the Periodic Table are arranged in order of ascending atomic number, starting with hydrogen (atomic number 1) at the top. The **atomic number** is the number of protons in the atom.

You can use the Periodic Table to find an element if you know its atomic number. Likewise, if you know where the element is in the Periodic Table you can find its atomic number.

## Isotopes

All atoms of a particular element have the same number of protons. (Different numbers of protons indicate atoms of different elements.) However, some elements have varieties that have different numbers of neutrons; these are called **isotopes**.

Isotopes have the same atomic number but a different mass number. The **mass number** is the total number of protons and neutrons in an atom.

**Mass number = Number of protons + Number of neutrons**

## HT More on Atomic Number and Isotopes

The table below shows some common elements and how the number of protons, neutrons and electrons can be calculated from the atomic number and mass number.

| Chemical symbol | Chemical name | Number of protons | Number of electrons | Number of neutrons |
|---|---|---|---|---|
| $^{1}_{1}\mathrm{H}$ | Hydrogen | 1 | 1 | 0 (1 − 1 = 0) |
| $^{3}_{1}\mathrm{H}$ | Hydrogen | 1 | 1 | 2 (3 − 1 = 2) |
| $^{4}_{2}\mathrm{He}$ | Helium | 2 | 2 | 2 (4 − 2 = 2) |
| $^{16}_{8}\mathrm{O}$ | Oxygen | 8 | 8 | 8 (16 − 8 = 8) |
| $^{23}_{11}\mathrm{Na}$ | Sodium | 11 | 11 | 12 (23 − 11 = 12) |

$^{3}_{1}\mathrm{H}$ is also called the hydrogen-3 isotope

## Electronic Structure

The electronic configuration tells us how the electrons are arranged around the nucleus.

Aluminium has the electronic structure 2.8.3. This means it has 3 occupied shells and a total of 13 electrons. The element with an electronic structure of 2.6 has 8 electrons; this will be the same as its atomic number and we can see that this is oxygen by looking it up on the Periodic Table.

- The electrons in an atom occupy the lowest available shells.
- The first energy level or shell can contain a maximum of only 2 electrons.
- The outer shell can hold a maximum of 8 electrons – an outer shell containing 8 electrons is known as a full outer shell.
- The electronic structure of an ion can be determined by working out the electronic structure of the atom and adding electrons for a negative ion or subtracting electrons for a positive ion.

HT

# The Modern Periodic Table – Electronic Structure of the First 20 Elements

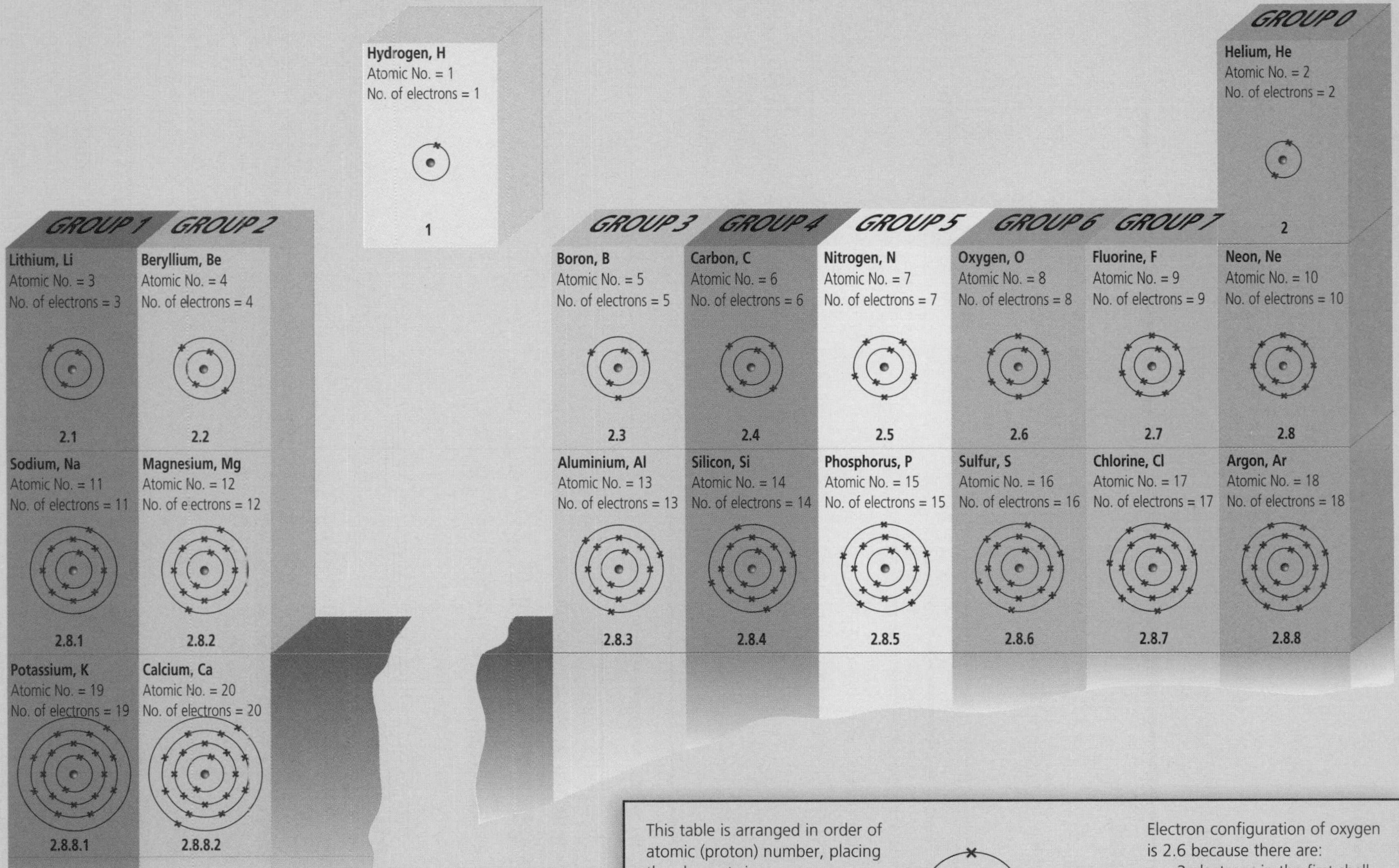

Hydrogen, H
Atomic No. = 1
No. of electrons = 1
1

| GROUP 1 | GROUP 2 | GROUP 3 | GROUP 4 | GROUP 5 | GROUP 6 | GROUP 7 | GROUP 0 |
|---|---|---|---|---|---|---|---|
| | | | | | | | **Helium, He** Atomic No. = 2 No. of electrons = 2 **2** |
| **Lithium, Li** Atomic No. = 3 No. of electrons = 3 **2.1** | **Beryllium, Be** Atomic No. = 4 No. of electrons = 4 **2.2** | **Boron, B** Atomic No. = 5 No. of electrons = 5 **2.3** | **Carbon, C** Atomic No. = 6 No. of electrons = 6 **2.4** | **Nitrogen, N** Atomic No. = 7 No. of electrons = 7 **2.5** | **Oxygen, O** Atomic No. = 8 No. of electrons = 8 **2.6** | **Fluorine, F** Atomic No. = 9 No. of electrons = 9 **2.7** | **Neon, Ne** Atomic No. = 10 No. of electrons = 10 **2.8** |
| **Sodium, Na** Atomic No. = 11 No. of electrons = 11 **2.8.1** | **Magnesium, Mg** Atomic No. = 12 No. of electrons = 12 **2.8.2** | **Aluminium, Al** Atomic No. = 13 No. of electrons = 13 **2.8.3** | **Silicon, Si** Atomic No. = 14 No. of electrons = 14 **2.8.4** | **Phosphorus, P** Atomic No. = 15 No. of electrons = 15 **2.8.5** | **Sulfur, S** Atomic No. = 16 No. of electrons = 16 **2.8.6** | **Chlorine, Cl** Atomic No. = 17 No. of electrons = 17 **2.8.7** | **Argon, Ar** Atomic No. = 18 No. of electrons = 18 **2.8.8** |
| **Potassium, K** Atomic No. = 19 No. of electrons = 19 **2.8.8.1** | **Calcium, Ca** Atomic No. = 20 No. of electrons = 20 **2.8.8.2** | | | | | | |

This table is arranged in order of atomic (proton) number, placing the elements in groups.

Elements in the same group have the same number of electrons in their highest occupied energy level (outer shell).

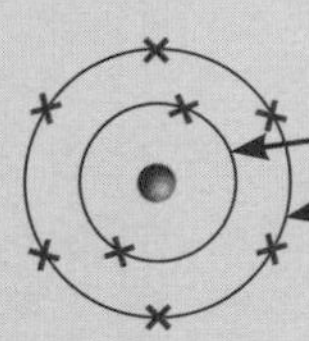

Electron configuration of oxygen is 2.6 because there are:
- 2 electrons in the first shell
- 6 electrons in the second shell.

# C4 Ionic Bonding

## Ions, Atoms and Molecules

An **uncharged** particle is either:

- an atom on its own, e.g. Na, Cl
- a molecule with two or more atoms bonded together, e.g. $Cl_2$, $CO_2$.

An **ion** is a charged atom or group of atoms, e.g. $Na^+$, $Cl^-$, $NH_4^+$, $SO_4^{2-}$.

## Forming Ions

A **positive ion** is formed when an atom (usually a metal) or group of atoms loses one or more electrons.

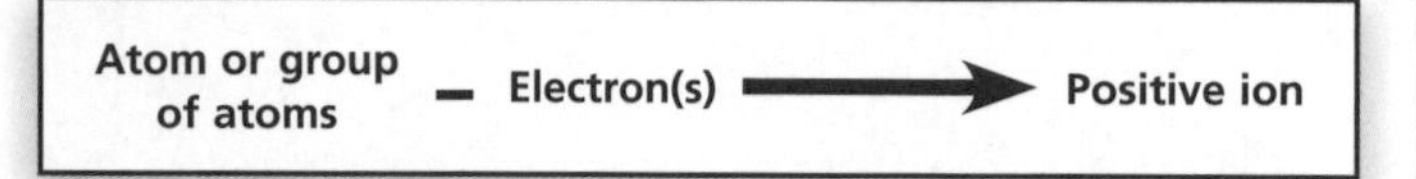


- Losing 1 electron makes a + ion, e.g. $Na^+$.
- Losing 2 electrons makes a 2+ ion, e.g. $Mg^{2+}$.

A **negative ion** is formed when an atom (usually a non-metal) or group of atoms gains one or more electrons.

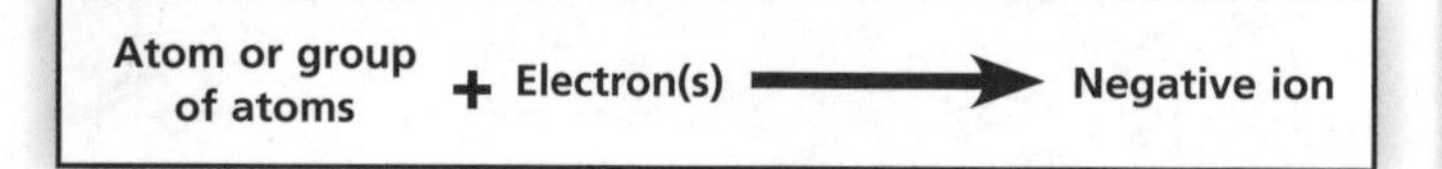


- Gaining 1 electron makes a – ion, e.g. $Cl^-$.
- Gaining 2 electrons makes a 2– ion, e.g. $O^{2-}$.

Atoms form ions to gain a full outer shell of electrons; this is a very stable structure.

## Ionic Bonding

A **metal** and a **non-metal** combine by **transferring electrons**. The metal atoms transfer electrons to become positive ions and the non-metal atoms receive electrons to become negative ions.

The positive ions and negative ions are then **attracted** to each other. Two compounds that are **bonded** by the attraction of oppositely charged ions are sodium chloride and magnesium oxide.

**Sodium chloride** has a high melting point. It does not conduct electricity when it is solid. However, it can dissolve in water, and the solution produced can conduct electricity. It is also able to conduct electricity when it is molten.

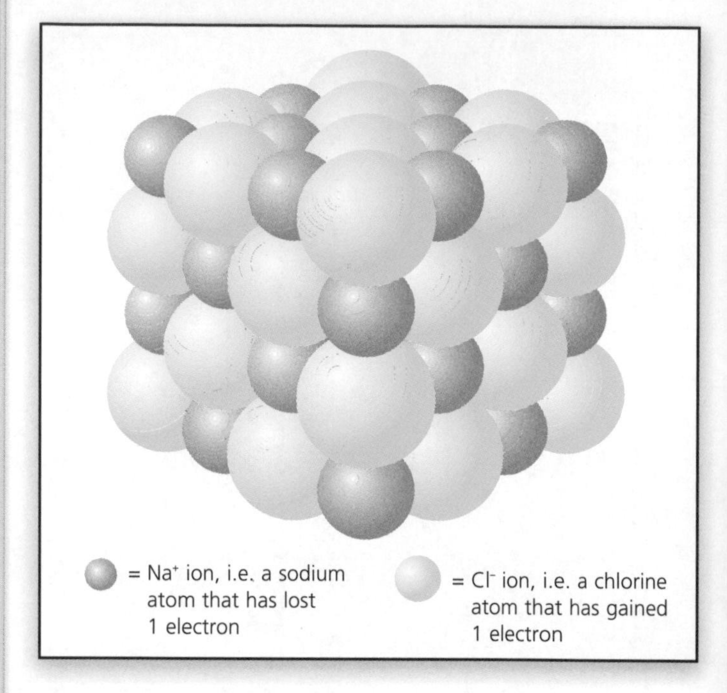


**Magnesium oxide** has an even higher melting point. It does not conduct electricity when it is solid. However, it can conduct electricity when it is molten.

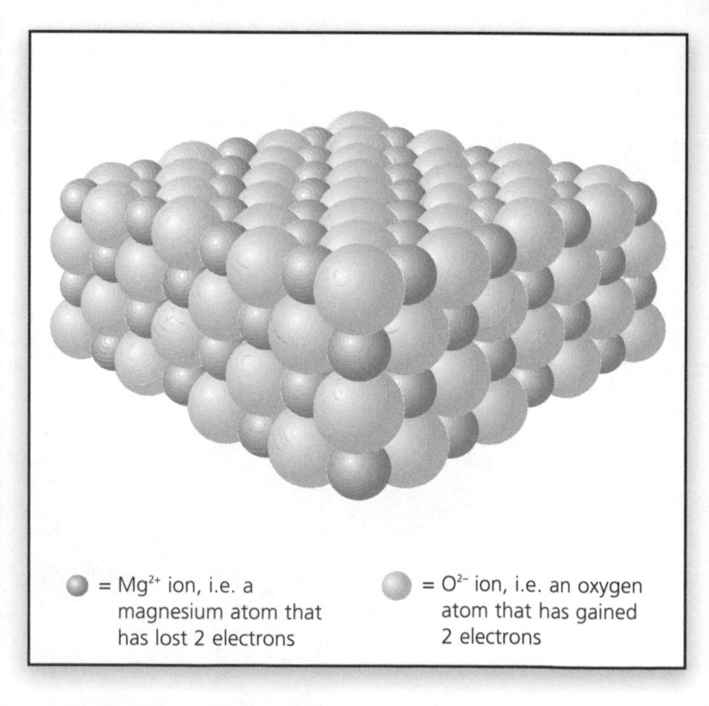


Sodium chloride (NaCl) and magnesium oxide (MgO) form **giant ionic lattices** in which positive ions and negative ions are **electrostatically** attracted to each other.

## Structure and Physical Properties of Sodium Chloride and Magnesium Oxide

The strong attraction between oppositely charged ions has to be overcome for them to melt and this results in sodium chloride and magnesium oxide having high melting points.

The attraction is stronger in magnesium oxide than in sodium chloride and so magnesium oxide has a higher melting point.

They conduct electricity when molten or in solution because the charged ions are free to move about.

When they are solid, the ions are held in place and cannot move about, which means they do not conduct electricity.

## Formulae of Ionic Compounds

All ionic compounds are neutral substances that have equal charges on the positive ion(s) and negative ion(s).

The table below shows how ions with different charges combine to form ionic compounds.

## The Ionic Bond

When a **metal** and a **non-metal** combine, electrons are transferred from one atom to the other to form **ions**. Each ion then has a **complete outer shell** (a **stable octet**).

### Example 1

The sodium atom has 1 electron in its outer shell which is transferred to the chlorine atom to give them both 8 electrons in their outer shell.

The atoms become ions ($Na^+$ and $Cl^-$) and the compound formed is sodium chloride, NaCl.

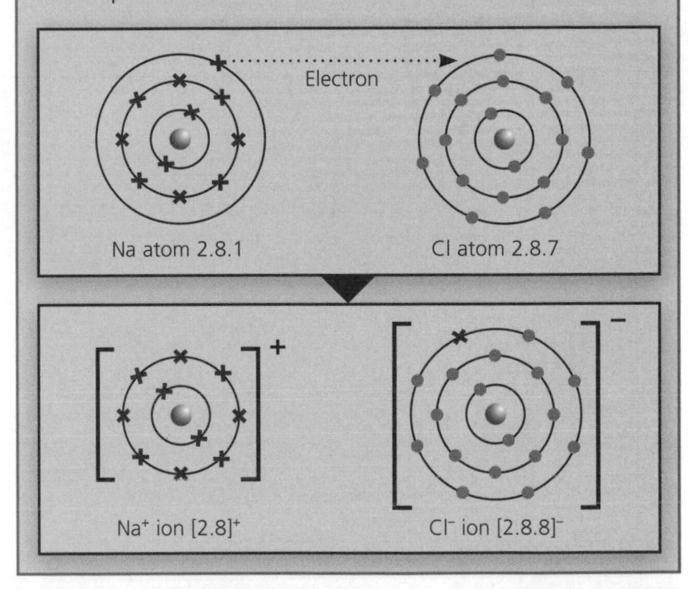


| Positive Ions | Negative Ions: 1– e.g. $Cl^-$, $OH^-$ | | Negative Ions: 2– e.g. $SO_4^{2-}$, $O^{2-}$ | |
|---|---|---|---|---|
| 1+ e.g. $K^+$, $Na^+$ | KCl: 1+ / 1– | NaOH: 1+ / 1– | $K_2SO_4$: 2 × 1+ = 2+ / 2– | $Na_2O$: 2 × 1+ = 2+ / 2– |
| 2+ e.g. $Mg^{2+}$, $Cu^{2+}$ | $MgCl_2$: 2+ / 2 × 1– = 2– | $Cu(OH)_2$: 2+ / 2 × 1– = 2– | $MgSO_4$: 2+ / 2– | CuO: 2+ / 2– |
| 3+ e.g. $Al^{3+}$, $Fe^{3+}$ | $AlCl_3$: 3+ / 3 × 1– = 3– | $Fe(OH)_3$: 3+ / 3 × 1– = 3– | $Al_2(SO_4)_3$: 2 × 3+ = 6+ / 3 × 2– = 6– | $Fe_2O_3$: 2 × 3+ = 6+ / 3 × 2– = 6– |

## HT The Ionic Bond (cont)

### Example 2

The magnesium atom has 2 electrons in its outer shell which are transferred to the oxygen atom to give them both 8 electrons in their outer shell.

The atoms become ions ($Mg^{2+}$ and $O^{2-}$) and the compound formed is magnesium oxide, MgO.

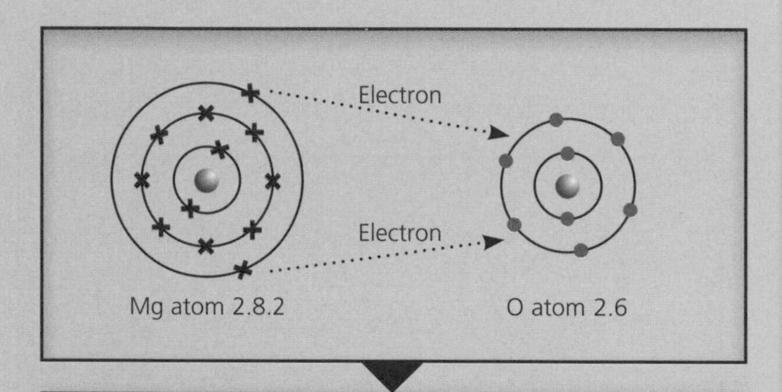


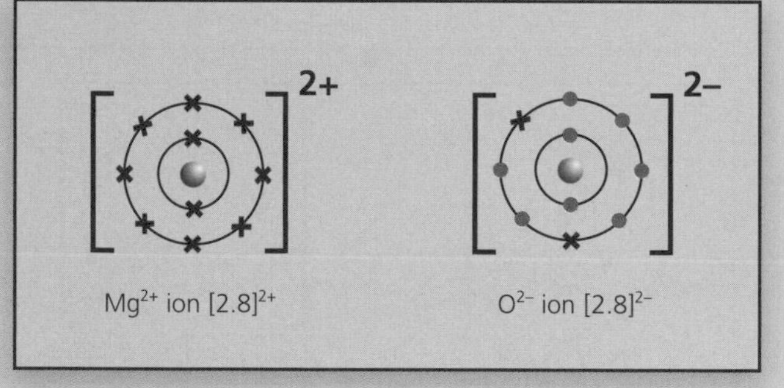


### Example 3

**magnesium + chlorine → magnesium chloride**

The magnesium atom has 2 electrons in its outer shell. A chlorine atom only needs 1 electron, therefore, 2 chlorine atoms are needed.

The atoms become ions ($Mg^{2+}$, $Cl^-$ and $Cl^-$) and the compound formed is magnesium chloride, $MgCl_2$.

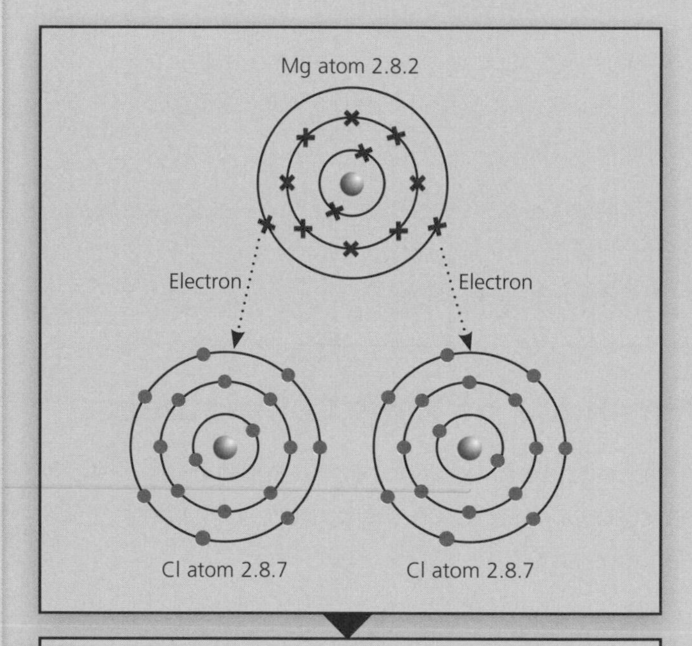


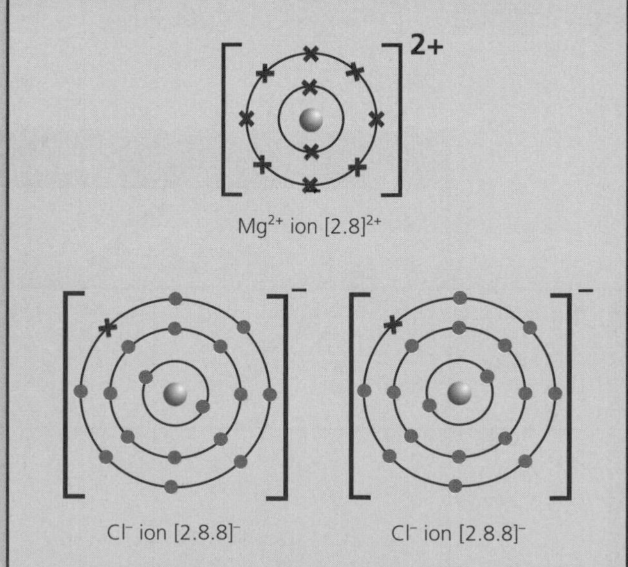

# The Periodic Table and Covalent Bonding

## The Periodic Table

Elements are the building blocks of all materials. The 100 or so elements are arranged in order of **ascending atomic number**, and then arranged in rows (periods) so that elements with similar properties are in the same column (group).

This forms the basis of the **Periodic Table**. More than three-quarters of the elements are metals. The others are non-metals.

The Periodic Table is more than just a list of elements. It has a pattern that is linked to the elements' properties.

**Johann Döbereiner** was the first scientist to try to arrange the elements in a pattern. He linked elements that had similar properties and found that they made **triads**, i.e. groups of three elements, e.g. chlorine, bromine and iodine. He noticed that the atomic weight of the triad's second element was the average of the other two. Although some elements could not be put into a triad, he published his idea in 1817.

In 1864 **John Newlands** arranged the elements by increasing atomic weight. He noticed that every eighth element had similar properties and coincided with the triads.

Other scientists did not accept his **Law of Octaves** because some elements had not yet been discovered and some did not fit the pattern.

**Dimitri Mendeleev** refined Newlands' ideas by putting all the known data for each element onto cards which he arranged in groups and periods in weight order based on their properties. If the weight order contradicted the properties of two elements he swapped them around to maintain his patterns. He left gaps for undiscovered elements and predicted their properties.

Mendeleev's ideas were published in 1869 but were not accepted. A few years later, a missing element was discovered. The properties matched his predictions and the Periodic Table was established.

HT Mendeleev's periodic table became widely accepted after the discovery of gallium, which had the properties he had predicted. The later discovery of electrons and the structure of atoms helped to explain Mendeleev's patterns. The number of electrons in each shell matched the rows in his periodic table.

Mendeleev had ignored atomic weight order for some elements and swapped them to match their properties. He was shown to be correct when the elements were arranged in atomic number order as in the modern Periodic Table.

The atomic weight order was incorrect due to the existence of heavier isotopes of elements with the lower atomic number (e.g. tellurium – atomic number 52, $A_r$ 128; iodine – atomic number 53, $A_r$ 127).

## Groups and Periods

A vertical column of elements in the Periodic Table is called a **group**, e.g. Group 1 contains lithium (Li), sodium (Na) and potassium (K), among others. Elements in the same group have similar chemical properties. This is because they have the **same number of electrons in their outer shell**. This number also coincides with the group number, e.g. Group 1 elements have 1 electron in their outer shell and Group 7 elements have 7 electrons in their outer shell.

A **horizontal row** of elements in the Periodic Table is called a **period**, e.g. lithium (Li), carbon (C) and neon (Ne) are all elements in the second period.

The period to which the element belongs corresponds to the **number of shells of electrons** it has, e.g. sodium (Na), aluminium (Al) and chlorine (Cl) all have three shells of electrons so they are found in the third period.

# C4 The Periodic Table and Covalent Bonding

## Bonding

A **molecule** is two or more atoms bonded together. There are two types of bonding:

- **Covalent bonding** – non-metals combine by sharing electrons.
- **Ionic bonding** – metals and non-metals combine by transferring electrons.

Two examples of covalently bonded molecules are **water** and **carbon dioxide**. Water, $H_2O$, is a liquid with a low melting point, which does not conduct electricity.

Carbon dioxide, $CO_2$, is a gas with a low melting point, which does not conduct electricity.

Water and carbon dioxide are simple molecules. They have low melting points because the forces between the molecules (intermoleculer forces) are weak, so only a small amount of energy is needed to overcome them.

## HT Representing Molecules

The following are all examples of **covalently bonded molecules**. You need to be familiar with how they are formed:

**Water ($H_2O$)** – the outer shells of the hydrogen and oxygen atoms overlap, and the oxygen atom shares a pair of electrons with each hydrogen atom to form a water molecule:

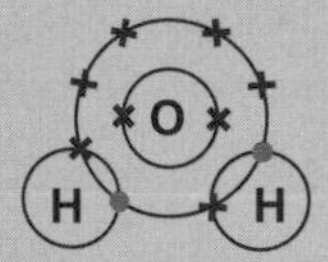


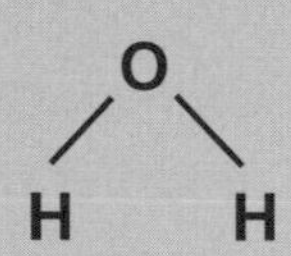


**Hydrogen ($H_2$)** – the two hydrogen atoms share a pair of electrons:

H – H

**Methane ($CH_4$)** – the carbon atom shares a pair of electrons with each hydrogen atom:

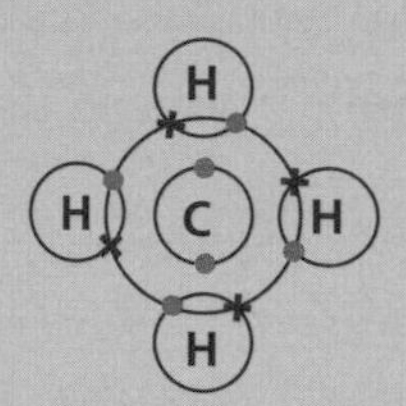


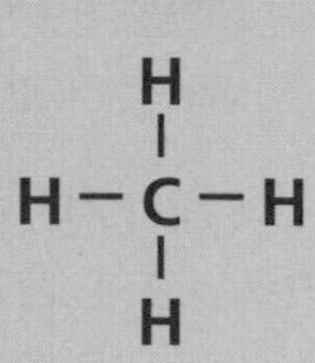


**Chlorine ($Cl_2$)** – the two chlorine atoms share a pair of electrons:

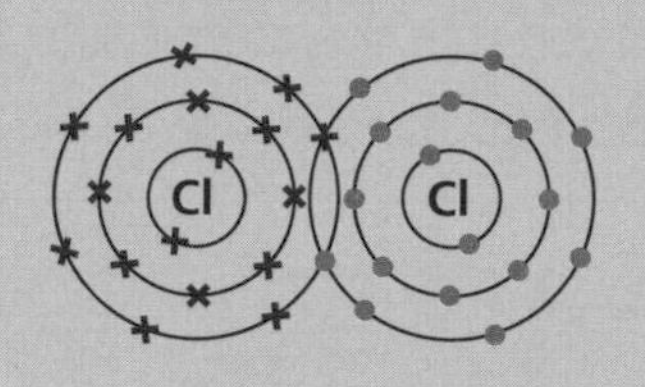


Cl – Cl

**Carbon dioxide ($CO_2$)** – the outer shells of the carbon and oxygen atoms overlap and the carbon atom shares two pairs of electrons with each oxygen atom to form a double covalently bonded molecule:

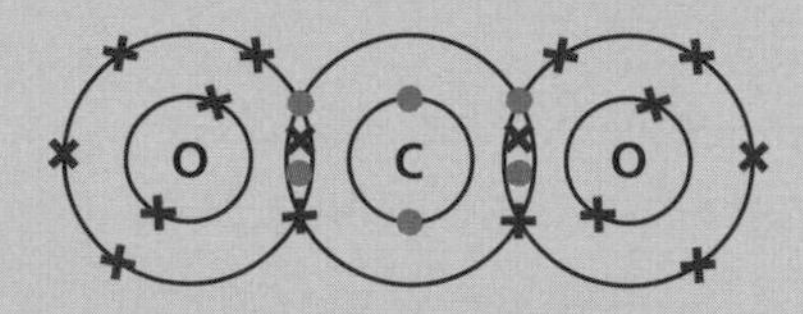


O = C = O

### Properties of Simple Covalently Bonded Molecules

The bond between two atoms in a simple covalently bonded molecule (e.g. water or carbon dioxide) is very strong.

The **intermolecular forces** of attraction *between* molecules are weak. This results in them having low melting points. They do not conduct electricity as they do not have any free electrons.

## Group 1 – The Alkali Metals

The **alkali metals** occupy the first vertical column (**Group 1**) at the left-hand side of the Periodic Table. The first three elements in the group are lithium, sodium and potassium. They all have one electron in their outer shell which means they have similar properties. Alkali metals are stored under oil because they react with air and react vigorously with water.

## Flame Tests

Lithium, sodium and potassium compounds can be recognised by the colours they produce in a **flame test**. The method used is explained below:

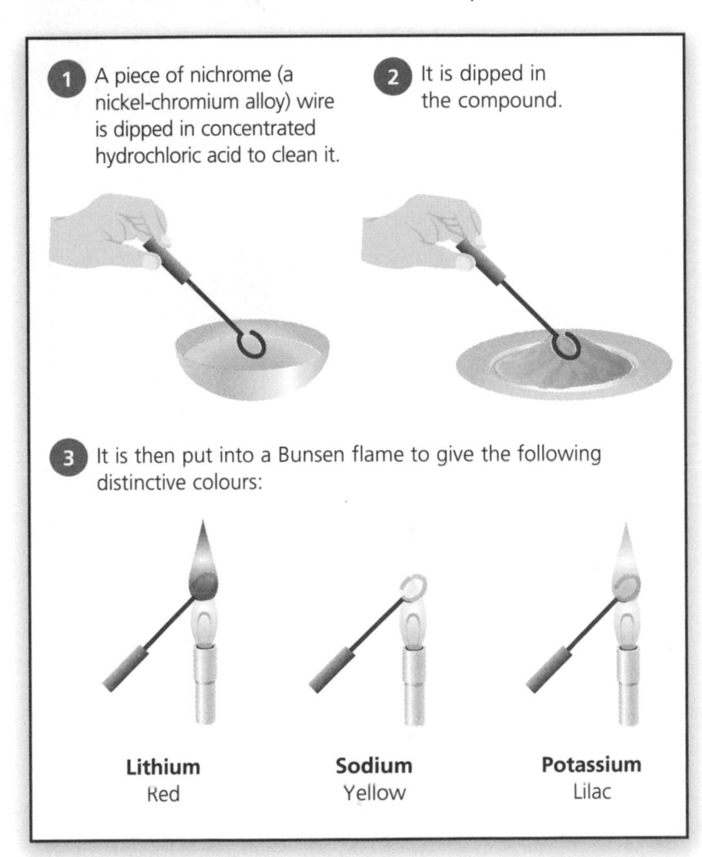


## Reacting Alkali Metals with Water

Alkali metals react with water to produce hydrogen and a **hydroxide.** Alkali metal hydroxides are soluble and form **alkaline solutions**.

As we go **down** the group, the alkali metals become more reactive and so they react more vigorously with water. They all float and some may melt and produce hydrogen gas (which may ignite).

Lithium reacts gently with water, sodium reacts more aggressively and potassium reacts so aggressively that it melts and burns with a lilac flame.

The diagram below shows what happens when a small piece of potassium is dropped into water:

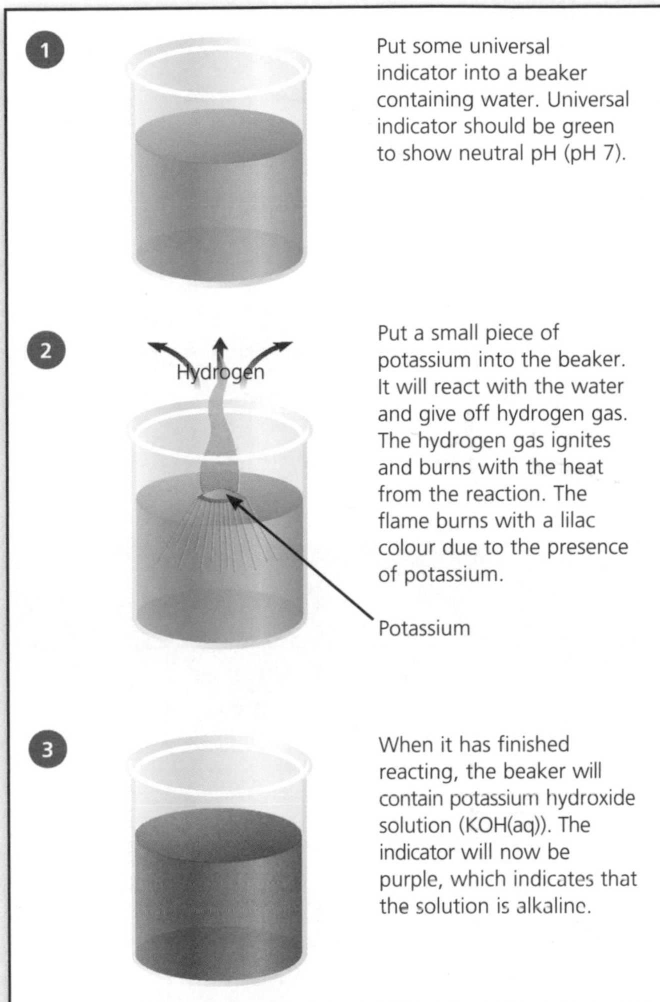


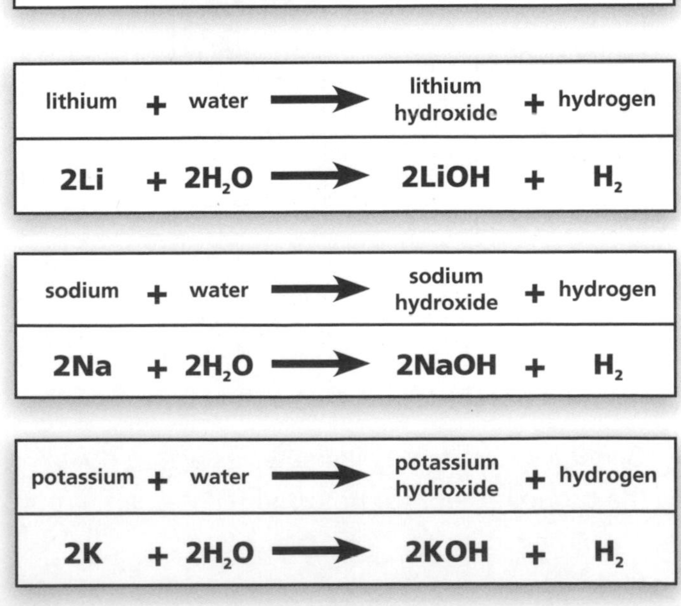

| lithium | + | water | → | lithium hydroxide | + | hydrogen |
|---|---|---|---|---|---|---|
| $2Li$ | + | $2H_2O$ | → | $2LiOH$ | + | $H_2$ |

| sodium | + | water | → | sodium hydroxide | + | hydrogen |
|---|---|---|---|---|---|---|
| $2Na$ | + | $2H_2O$ | → | $2NaOH$ | + | $H_2$ |

| potassium | + | water | → | potassium hydroxide | + | hydrogen |
|---|---|---|---|---|---|---|
| $2K$ | + | $2H_2O$ | → | $2KOH$ | + | $H_2$ |

## Properties of the Alkali Metals

Rubidium is the fourth element in Group 1. Rubidium, like the other elements in the group, reacts with water. The reaction would be:

- very fast
- exothermic
- violent (if it is carried out in a glass beaker, the beaker may shatter).

Caesium (the fifth element in the group) is even more reactive and will react very violently with water.

HT Even though the alkali metals have similar chemical properties, their physical properties alter as we go down the group.

The table below shows their melting and boiling points and their densities:

| Element | Melting Point (°C) | Boiling Point (°C) | Density ($g/cm^3$) |
|---|---|---|---|
| Lithium, Li | 180 | 1340 | 0.53 |
| Sodium, Na | 98 | 883 | 0.97 |
| Potassium, K | 64 | 760 | 0.86 |
| Rubidium, Rb | 39 | 688 | 1.53 |
| Caesium, Cs | 29 | 671 | 1.90 |

The melting points and boiling points of alkali metals decrease going down the group.

Caesium has the lowest melting and boiling points.

Generally, the density increases as we go down the group (except for potassium). Caesium has the largest density.

## Trends in Group 1

Alkali metals have similar properties because when they react, an atom loses one electron to form a **positive ion** with a **stable electronic structure**, i.e. it has a full outer shell of electrons.

The alkali metals become more reactive as we go down the group because the outer shell gets further away from the influence of the nucleus, making it easier for an atom to lose an electron from this shell.

**Oxidation** involves the loss of electrons from an atom. Examples are shown below:

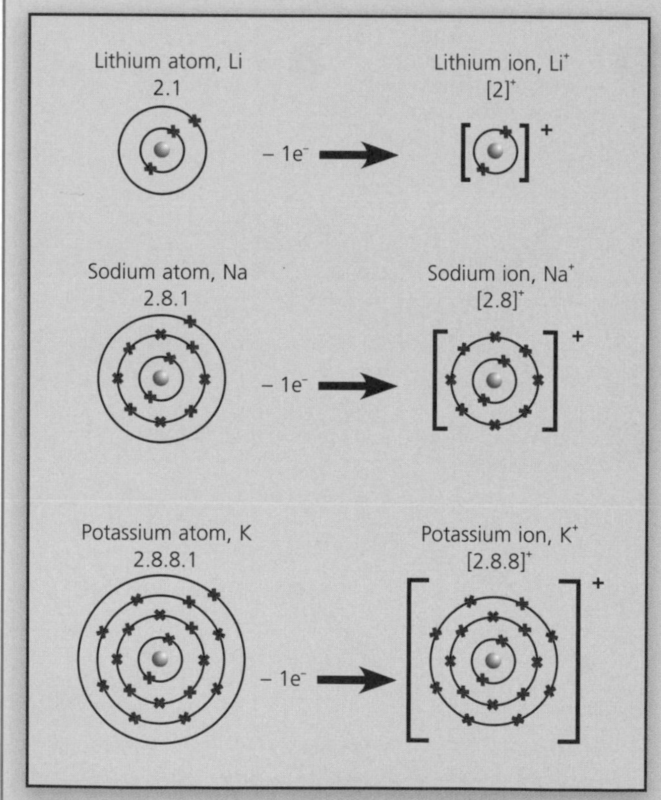


The equations for the formation of the Group 1 metal ions are usually written as follows:

$$Li - e^- \longrightarrow Li^+$$

$$Na - e^- \longrightarrow Na^+$$

$$K - e^- \longrightarrow K^+$$

## Group 7 – The Halogens

There are five non-metals in Group 7 and they are known as the **halogens**. They all have seven electrons in their outer shell which means that they have similar chemical properties. You need to know about four of the halogens: **fluorine**, **chlorine**, **bromine** and **iodine**.

At room temperature:

- chlorine is a green gas
- bromine is an orange liquid
- iodine is a grey solid.

Iodine is used as an antiseptic to sterilise wounds.

Chlorine, the most commonly used halogen, is used to sterilise water and to make pesticides and plastics.

## Reactions with Alkali Metals

Halogens react vigorously with alkali metals to form metal halides.

Use these examples to work out the product and write the word equation for any combination of halogen and alkali metal:

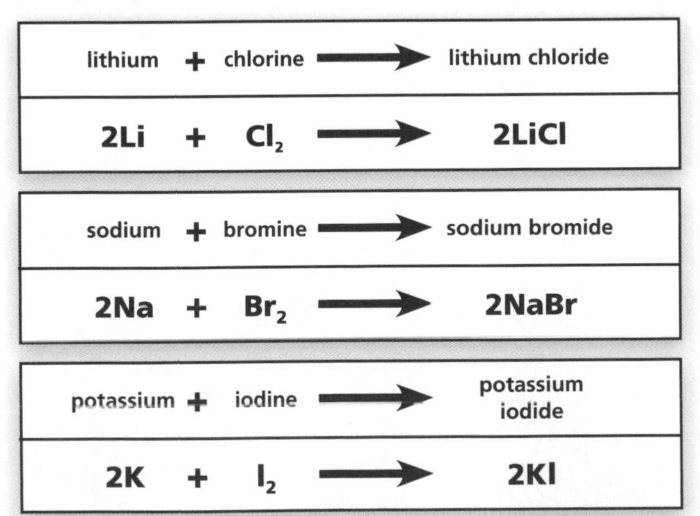

| lithium + chlorine → lithium chloride |
|---|
| $2Li + Cl_2 \rightarrow 2LiCl$ |

| sodium + bromine → sodium bromide |
|---|
| $2Na + Br_2 \rightarrow 2NaBr$ |

| potassium + iodine → potassium iodide |
|---|
| $2K + I_2 \rightarrow 2KI$ |

## Displacement Reactions

As we go down the group, the halogens become **less reactive**, and their melting and boiling points increase.

Fluorine is therefore the most reactive halogen and iodine is the least reactive.

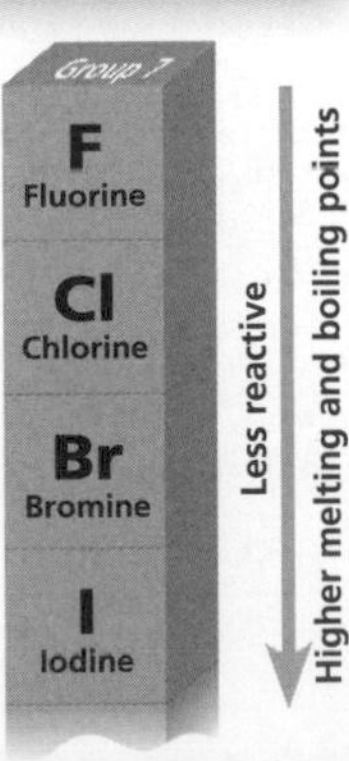


A more reactive halogen will displace a less reactive halogen from an aqueous solution of its metal **halide**, i.e. chlorine will displace bromides and iodides, and bromine will displace iodides.

If chlorine gas is passed through an aqueous solution of potassium bromide, bromine is formed due to the displacement reaction taking place:

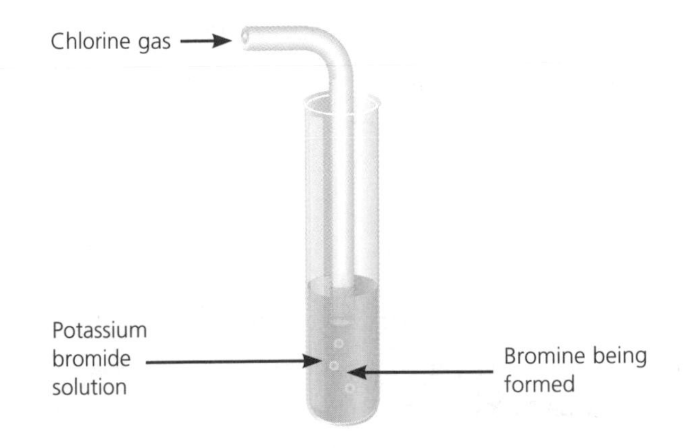


The table below shows the products of the reactions between halogens and aqueous solutions of halide salts:

| | Potassium chloride | Potassium bromide | Potassium iodide |
|---|---|---|---|
| Chlorine $Cl_2$ | X | Potassium chloride + bromine | Potassium chloride + iodine |
| Bromine $Br_2$ | No reaction | X | Potassium bromide + iodine |
| Iodine $I_2$ | No reaction | No reaction | X |

X = No experiment conducted.

| potassium bromide + chlorine → potassium chloride + bromine |
|---|
| $2KBr + Cl_2 \rightarrow 2KCl + Br_2$ |

| potassium iodide + chlorine → potassium chloride + iodine |
|---|
| $2KI + Cl_2 \rightarrow 2KCl + I_2$ |

| potassium iodide + bromine → potassium bromide + iodine |
|---|
| $2KI + Br_2 \rightarrow 2KBr + I_2$ |

## Properties of the Halogens

Fluorine is the first element in Group 7, and is the most reactive element in the group. It will displace all of the other halogens from an aqueous solution of their metal halides.

Astatine is the fifth element in Group 7. It is a semi-metallic, radioactive element and only very small amounts of it exist naturally. Even so, it is the least reactive of the halogens and, theoretically, it would be unable to displace any of the other halogens from an aqueous solution of their metal halides.

The **physical properties** of the halogens alter as we go down the group. The table below shows their melting and boiling points and their densities:

| Element | Melting Point (°C) | Boiling Point (°C) | Density (g/cm³) |
|---|---|---|---|
| Fluorine | –220 | –188 | 0.0016 |
| Chlorine | –101 | –34 | 0.003 |
| Bromine | –7 | 59 | 3.12 |
| Iodine | 114 | 184 | 4.95 |
| Astatine | 302 (estimated) | 337 (estimated) | about 7 (estimated) |

The melting points and boiling points of the halogens increase going down the group. The density also increases going down the group.

Astatine is estimated to have the highest melting point, boiling point and density. The figure for density in the table is estimated by looking at the trend of differences in density between each member of the group and adding that difference to the value for iodine. Accurate figures for astatine are not known because astatine is a very unstable element.

## Trends in Group 7

The halogens have similar properties because when they react, an atom **gains** one electron to form a **negative ion** with a stable electronic structure, i.e. it has a full outer shell of electrons. **Reduction** involves the gain of electrons by an atom. Examples are shown below:

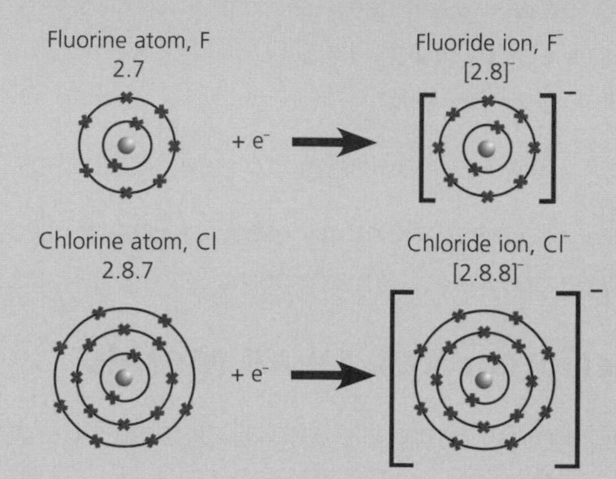


Fluorine is the most reactive because it is easiest for it to gain the extra electron.

The halogens become less reactive as we go down the group because the outer shell gets further away from the influence of the nucleus, making it harder for an atom to gain an electron.

The equations for the formation of the halide ions from halogen molecules are usually written as follows:

$$F_2 + 2e^- \rightarrow 2F^-$$

$$Cl_2 + 2e^- \rightarrow 2Cl^-$$

You can decide whether a reaction is an example of oxidation or reduction by looking at its equation, such as the ones above. If electrons are added, then it is a reduction reaction, and if electrons are taken away it is an oxidation reaction.

An easy way to remember the definitions of oxidation and reduction is **OIL RIG:**

- **O**xidation **I**s **L**oss of electrons
- **R**eduction **I**s **G**ain of electrons.

## The Transition Metals

In the centre of the Periodic Table, between Groups 2 and 3, is a block of metallic elements called the **transition metals**. This block includes **iron** (Fe), **copper** (Cu), **platinum** (Pt), **mercury** (Hg), **chromium** (Cr) and **zinc** (Zn).

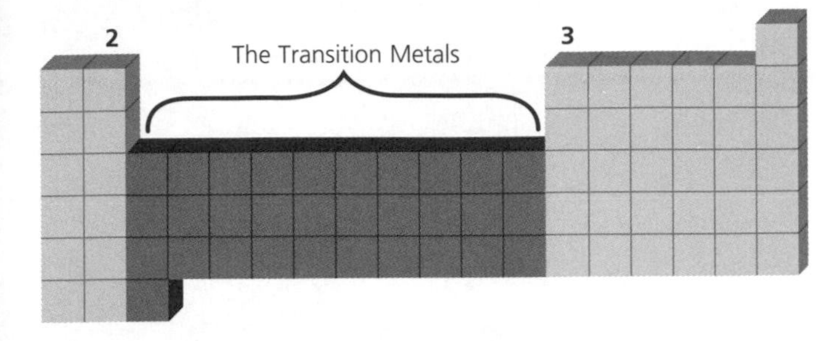


Transition metals have the typical properties of metals.

Compounds of transition metals are often **coloured**:

- copper compounds are blue
- iron(II) compounds are grey–green
- iron(III) compounds are orange–brown.

Many transition metals and their compounds can be used as **catalysts** in chemical reactions, for example:

- iron is used in the Haber process
- nickel is used in the manufacture of margarine.

## Thermal Decomposition of Transition Metal Carbonates

**Thermal decomposition** is a reaction in which a substance is broken down into simpler substances by heating. When transition metal carbonates are heated, a **colour change** occurs and they decompose to form a **metal oxide** and **carbon dioxide**. For example, if copper carbonate is heated, the blue–green copper carbonate decomposes into black copper oxide and carbon dioxide, which turns limewater milky:

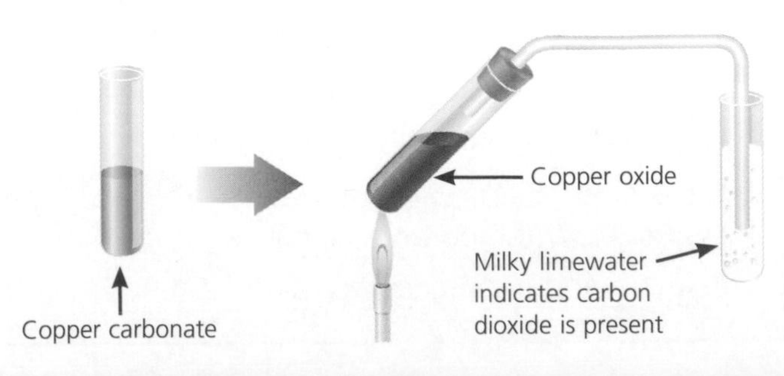


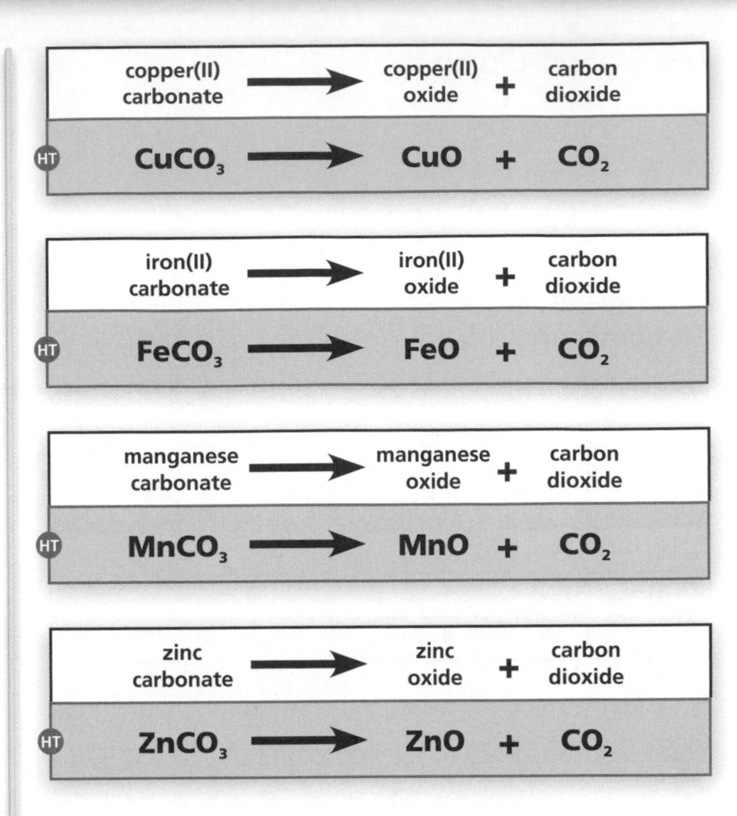


## Identifying Transition Metal Ions

Metal compounds in solution contain metal ions. Some of these form **precipitates** (insoluble solids) that come out of solution when sodium hydroxide solution is added to them. The following ions form coloured precipitates:

| Metal Ion | Colour of Precipitate | Ionic Symbol Equation |
|---|---|---|
| Copper(II), $Cu^{2+}$ | Blue | HT $Cu^{2+} + 2OH^- \rightarrow Cu(OH)_2$ |
| Iron(II), $Fe^{2+}$ | Grey–green | HT $Fe^{2+} + 2OH^- \rightarrow Fe(OH)_2$ |
| Iron(III), $Fe^{3+}$ | Orange–brown | HT $Fe^{3+} + 3OH^- \rightarrow Fe(OH)_3$ |

The reaction between the transition metal ions and sodium hydroxide solution is called a **precipitation** reaction.

# C4 Metal Structure and Properties

## Iron and Copper

Iron and copper are two transition metals that have many uses, for example:

- iron is used to make steel, which is used to make cars and girders because it is **very strong**
- copper is used to make electrical wiring as it is a **good conductor**. It is also used to make brass.

Look at the data about metals in this table:

| Metal | Relative Electrical Conductivity | Density ($g/cm^3$) | Relative Hardness |
|---|---|---|---|
| Zinc | 1.7 | 7.1 | 2.5 |
| Copper | 6.1 | 9.0 | 3.0 |
| Nickel | 1.4 | 8.9 | 4.0 |
| Cobalt | 1.7 | 8.9 | 5.0 |

You could be asked to pick out the hardest metal (i.e. cobalt) or explain why copper is used in electrical wires (i.e. it is the best electrical conductor).

## Physical Properties

Metals are very useful materials because of their properties. They:

- are **lustrous** (shiny), e.g. gold is used in jewellery
- are **hard** and have a **high density**, e.g. steel is used to make drill bits
- have high **tensile strength** (able to bear loads), e.g. steel is used to make girders
- have **high melting** and **boiling points**, e.g. tungsten is used to make light bulb filaments
- are **good conductors** of heat and electricity, e.g. copper is used to make saucepans and wiring.

## Metal Structure

Metal atoms are held together by metallic bonds and are packed very closely in a regular arrangement.

Metals have high melting and boiling points because lots of energy is needed to overcome the strong metallic bonds. As the metal atoms pack together, they build a structure of **crystals**.

HT Metal atoms are packed so close together that the outer electron shells **overlap** and form **metallic bonds**. The overlap allows electrons to move about freely. The structure can be described as closely packed metal ions in a 'sea' of **delocalised** (free) electrons.

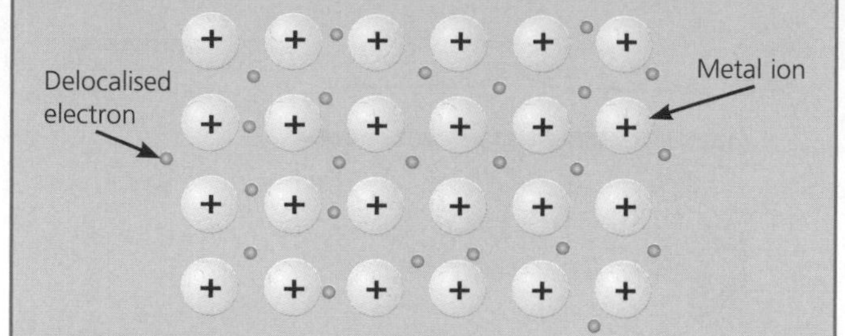


The free movement of the delocalised electrons allows the metal to conduct electricity (e.g. in wiring). The metal is held together by strong forces (the **electrostatic attraction** between the metal ions and the delocalised electrons). This is why many metals have high melting and boiling points.

## Superconductors

Metals are able to conduct electricity because the atoms are very close together and the electrons can move from atom to atom.

At low temperatures, some metals can become **superconductors**. A superconductor has very little, or no, resistance to the flow of electricity. Very low resistance is useful when you require:

- a powerful electromagnet, e.g. inside medical scanners
- very fast electronic circuits, e.g. in a supercomputer
- power transmission that does not lose energy.

HT The search is on to find a superconductor that will work at room temperature (20°C). The majority of superconductors currently in use operate at temperatures below -200°C. This very low temperature is costly to maintain and impractical for large-scale uses.

## Water

The four main **sources** of water are:

- rivers
- lakes
- reservoirs
- aquifiers (wells and bore holes).

**Example**

The pie chart shows the sources of water in Northern Ireland.

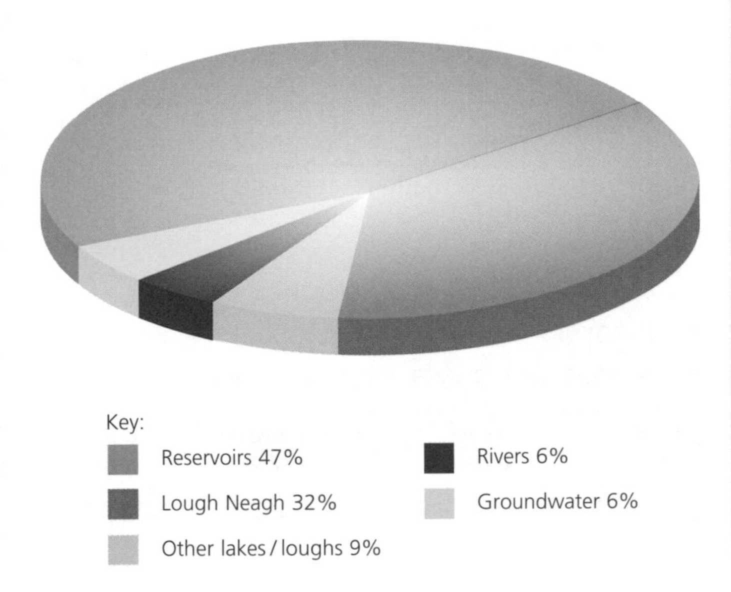


You will not be expected to remember this data, but you may be asked to interpret data like this. For example, using this data you could be asked to pick out the largest water source in Northern Ireland.

Water is an important resource for industry as well as being essential for drinking, washing and other jobs in the home. The chemical industry uses large amounts of water for **cooling**, as a **solvent** and as a cheap **raw material**.

In some parts of Britain the demand for water is higher than the supply, so it is important not to waste water.

## Water Treatment

Water has to be **treated** to **purify** it and make sure it is safe to drink before it reaches homes or factories. Untreated (raw) water can contain:

- insoluble particles
- pollutants
- microbes
- dissolved salts and minerals.

A typical treatment process is shown below:

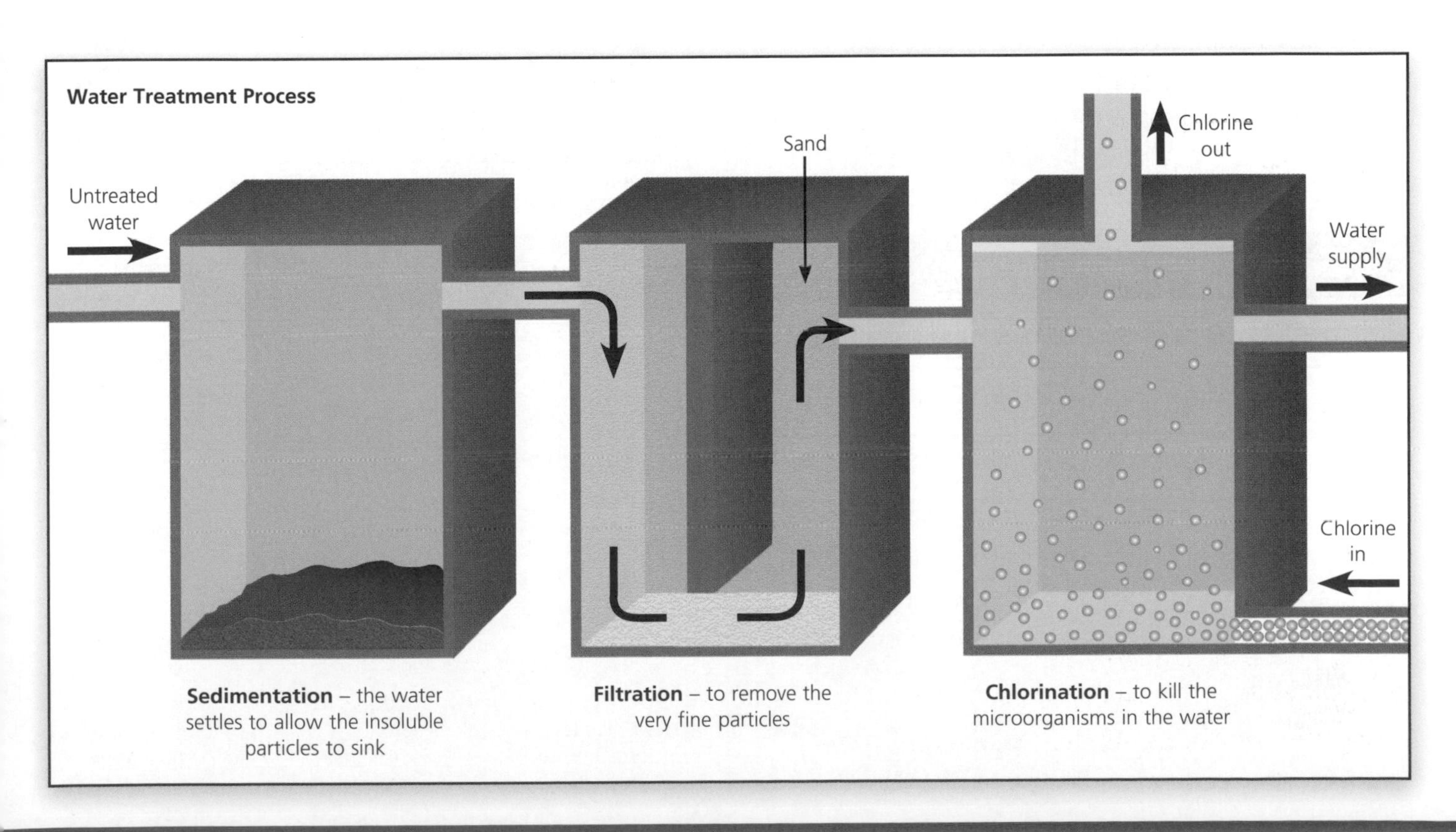

## Water Purification

HT

Tap water is not pure. It contains soluble materials that are not removed in the normal water treatment process. There is the possibility that some of the materials may be poisonous and, in that case, extra steps must be taken to remove them. Water must be distilled to make sure it is absolutely pure. This uses lots of energy, so it is very costly.

Huge amounts of expensive energy would be needed in order to distil sea water. And because sea water is quite corrosive, the equipment needed would be very costly. These factors make the cost of making drinking water out of sea water prohibitive, i.e. we are prevented from doing it because the cost is too high.

## Pollution in Water

The **pollutants** that may be found in water supplies include nitrates from the run-off of fertilisers, lead compounds from old pipes in the plumbing, and pesticides from spraying crops near to the water supply. These materials are more difficult to remove from the water.

The table below shows some pollutants and the maximum amounts permitted in drinking water.

| Pollutant | Maximum Amount Permitted |
|---|---|
| Nitrates | 50 parts in 1 000 000 000 parts water |
| Lead | 50 parts in 1 000 000 000 parts water |
| Pesticides | 0.5 parts in 1 000 000 000 parts water |

Again, you do not have to remember the data but you may be asked to pick out which pollutant has the smallest allowed concentration, or transfer the data onto a graph.

## Dissolved Ions

The **dissolved ions** of some salts are easy to identify as they will undergo **precipitation** reactions. A precipitation reaction occurs when an insoluble solid is made from mixing two solutions together.

Sulfates can be detected using barium chloride solution: a white precipitate of barium sulfate forms, as in the following example:

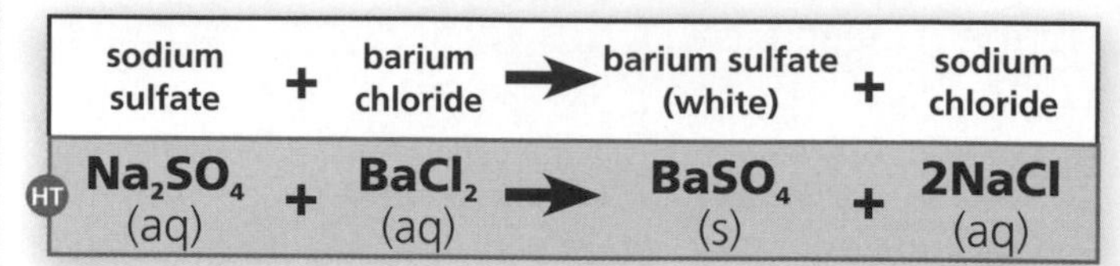

sodium sulfate + barium chloride → barium sulfate (white) + sodium chloride

HT $Na_2SO_4(aq) + BaCl_2(aq) \rightarrow BaSO_4(s) + 2NaCl(aq)$

Silver nitrate solution is used to detect halide ions. Halides are the ions made by the halogens (Group 7).

With silver nitrate:

- chlorides form a white precipitate
- bromides form a cream precipitate
- iodides form a pale yellow precipitate.

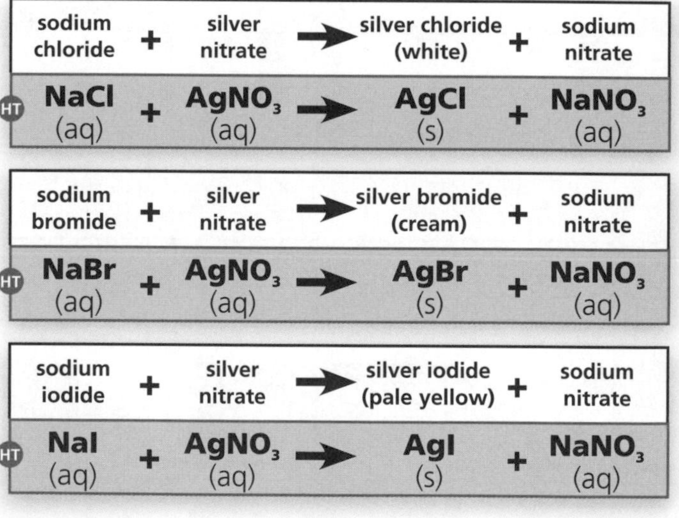

sodium chloride + silver nitrate → silver chloride (white) + sodium nitrate

HT $NaCl(aq) + AgNO_3(aq) \rightarrow AgCl(s) + NaNO_3(aq)$

sodium bromide + silver nitrate → silver bromide (cream) + sodium nitrate

HT $NaBr(aq) + AgNO_3(aq) \rightarrow AgBr(s) + NaNO_3(aq)$

sodium iodide + silver nitrate → silver iodide (pale yellow) + sodium nitrate

HT $NaI(aq) + AgNO_3(aq) \rightarrow AgI(s) + NaNO_3(aq)$

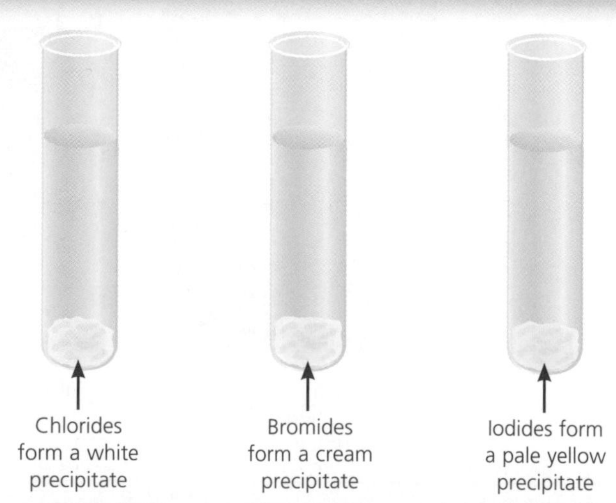

1. **a)** What are isotopes? **[2]**
   **b)** Calculate the number of protons, neutrons and electrons in $^{35}_{17}Cl$ **[3]**
   **c)** Work out the electronic structure of silicon. (Atomic number = 14) **[1]**
2. **a)** Explain how you make a $1^+$ ion ( e.g. $Na^+$) from a neutral atom. **[1]**
   **b)** Explain how you make a $2^-$ ion ( e.g. $O^{2-}$) from a neutral atom. **[1]**
3. **a)** Name the gas that is used to make plastics and pesticides and is also used to sterilise water. **[1]**
   **b)** Write the word equation to show the reaction of sodium iodide solution with the gas you named in part a). **[1]**
4. **a)** Gold and silver jewellery are lustrous. What does this mean? **[1]**
   **b)** Suggest one other property that is desirable in a metal used to make jewellery. **[1]**
   **c)** Brass is an alloy made of copper and zinc. Suggest why brass is a better material than copper for making musical instruments. **[1]**
5. **a)** Write the word equation to show the reaction of sodium iodide solution with chlorine gas. **[1]**
   **b)** Use the reaction to compare the reactivity of chlorine and iodine. **[1]**
6. **a)** Describe one benefit and one problem of using a superconductor. **[2]**
   **b)** How could a superconductor be useful in a supercomputer? **[1]**
7. **a)** Suggest which of the following is not a transition metal. Use a Periodic Table to help you. **[1]**

   **Copper** **Iron** **Zinc** **Silver** **Calcium** **Gold**

HT **b)** Complete and balance this equation:
$Fe^{2+} + OH^- \longrightarrow$ **[2]**

8. Explain why a silicon atom has no overall charge. **[1]**
9. Explain why ionic compounds such as sodium chloride have such high melting points and will conduct electricity when molten, but covalent compounds, such as methane, have low melting points and do not conduct electricity. **[6]**

   *The quality of written communication will be assessed in your answer to this question.*
10. **a)** Draw the electronic structure of a magnesium ion and a chloride ion (atomic numbers: Mg = 12; Cl = 17). **[2]**
    **b)** Use your answer to part (a) to work out the formula of magnesium chloride. **[1]**
    **c)** Write the ionic equation to show the formation of bromide ions ($Br^-$) from a bromine molecule ($Br_2$) and explain why the reaction is reduction. **[3]**
11. Draw a dot and cross diagram to show the covalent bonding in methane, $CH_4$. **[2]**
12. Calculate the number of protons, neutrons and electrons in $^{16}_{8}O^{2-}$ **[3]**
13. **a)** Draw the electronic structure of an atom with an atomic number of 11 **[1]**
    **b)** Draw the electronic structure of the positive ion formed by the atom in part a). **[1]**

# P3 Speed

**P3: Forces for Transport**

This module looks at:

- Speed, distance and time, alongside transport and road safety.
- Acceleration and how to measure it.
- Thinking, braking and stopping distances when driving.
- Work done, power and energy in relation to car use.
- Transport and energy, including the use of fossil fuels and the use of biofuels and solar power.
- How energy is absorbed during collisions, and car safety features.
- Falling objects and the balance of forces.
- Theme park rides and their forces and energy, including gravity and potential energy and kinetic energy.

## Speed

One way to describe the movement of an object is by measuring its **speed**, i.e. how fast it is moving.

A faster object will travel a longer distance in a given time. Or a faster object will take less time to travel a certain distance.

Speed is measured in:

- **metres per second (m/s)**
- **kilometres per hour (km/h)**
- **miles per hour (mph).**

For example, the cyclist in the diagram below travels a distance of 8 metres every second, so his speed is 8m/s.

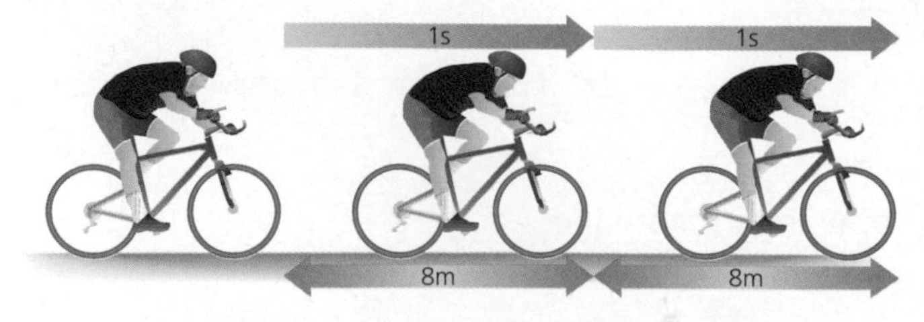


To work out the speed of any moving object two things must be known:

- the **distance** it travels (which can be measured using a measuring tape or a trundle wheel)
- the **time taken** to travel that distance (which can be measured using a stopclock).

The speed of an object can be calculated using the equation:

$$\textbf{Speed}\ (\text{m/s}) = \frac{\textbf{Distance travelled (m)}}{\textbf{Time taken}\ (\text{s})}$$

where $v$ is speed

$d$

$v \times t$

The faster the speed of an object, the shorter the time it takes to travel a particular distance.

### Example 1

Calculate the speed of a cyclist who travels 2400m in 5 minutes. Use the formula:

$$\text{Speed} = \frac{\text{Distance}}{\text{Time taken}}$$

$$= \frac{2400\text{m}}{300\text{s}} = \mathbf{8m/s}$$

You need to be able to rearrange the speed formula in order to calculate either distance or time taken.

### Example 2

A car is travelling at a constant speed of 80km/h. Calculate the distance it travels in 90 minutes. Rearrange the formula:

$$\text{Distance} = \text{Speed} \times \text{Time taken}$$

$$= 80\text{km/h} \times 1.5\text{h} = \mathbf{120km}$$

### Example 3

Calculate the time it takes a motorcyclist to travel a distance of 120km at 50km/h. Rearrange the formula:

$$\text{Time taken} = \frac{\text{Distance}}{\text{Speed}}$$

$$= \frac{120\text{km}}{50\text{km/h}} = 2.4\text{h} = \mathbf{2h\ 24min}$$

## Distance and Average Speed

For an object that is changing speed (uniformly) it is possible to calculate the distance it travels by using the average speed:

$$\text{Distance} = \text{Average speed} \times \text{Time}$$

$$\text{Distance} = \frac{(u + v)}{2} \times t$$

where $u$ is the starting speed and $v$ is the final speed

**Example**

A bike travelling at 2 m/s increases in speed to 10 m/s in 20 seconds. How far does it travel in this time?

$$\text{Distance} = \text{Average speed} \times \text{Time}$$
$$= \frac{(10 + 2)}{2} \times 20$$
$$= 6 \times 20 = \mathbf{120m}$$

**Speed Cameras**

Speed cameras take two pictures of a vehicle, one a certain amount of time after the other.

The position of the vehicle in relation to the distance markings on the road in the two pictures are used to calculate the speed of the vehicle, using the following formula:

$$\textbf{Speed of car} = \frac{\textbf{Distance travelled between pictures}}{\textbf{Time between first and second picture}}$$

**Average-speed cameras** are often used on motorways. A camera takes a photo of the car number plate and a second camera takes a photo of the same car later. The time between the photos and the distance between the cameras is used to work out the average speed.

## Distance–Time Graphs

The slope of a **distance–time graph** represents the speed of an object: the steeper the slope, the faster the speed. The $y$-axis shows the distance from a fixed point (0, 0), not the total distance travelled.

If an object (e.g. a person) is standing 10m from point (0, 0) and is not moving, the distance–time graph would look like this:

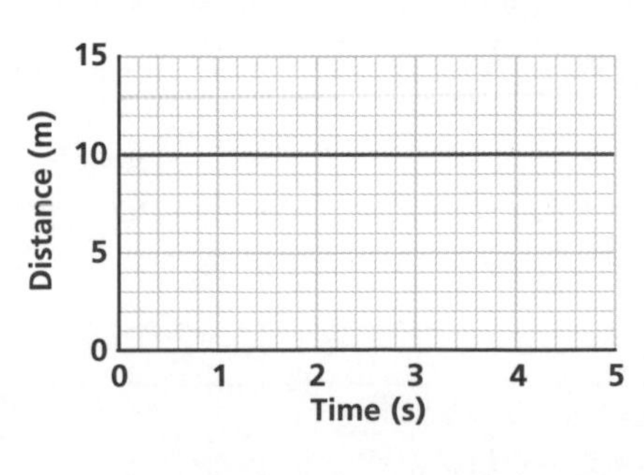

If the person starts at point (0, 0) and moves at a constant speed of 2m/s, the graph would look like this:

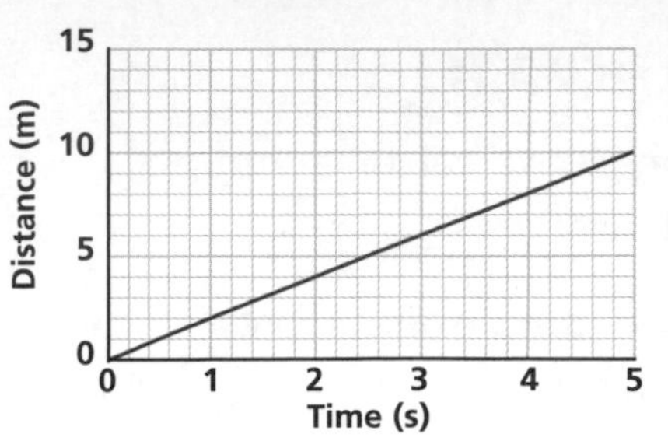


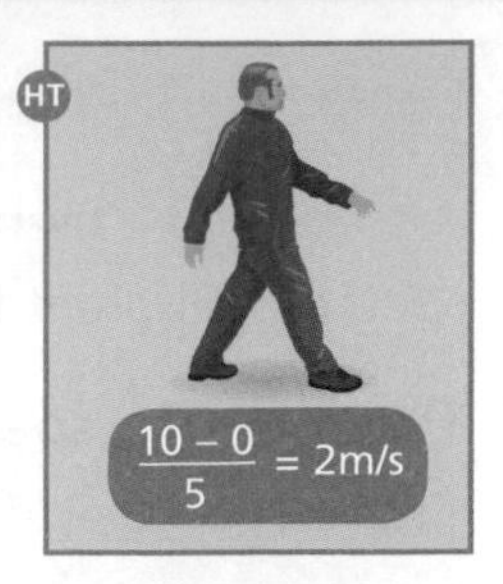


If the person starts at point (0, 0) and moves at a greater constant speed of 3m/s, the graph would look like this:

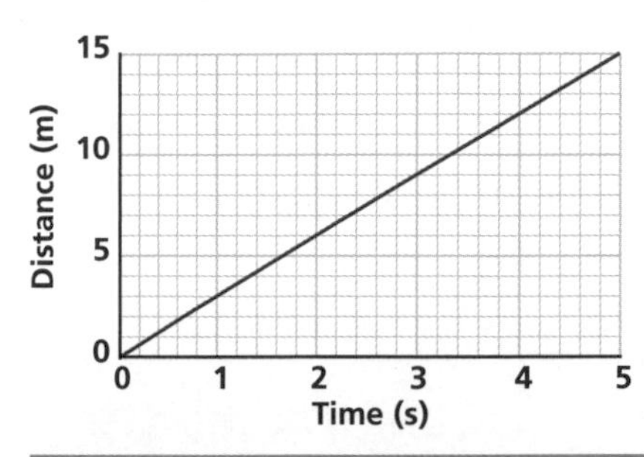

HT The speed of an object can be calculated by working out the gradient of a distance–time graph: the steeper the gradient, the faster the speed. Take any point on the graph and read off the distance travelled for that part of the journey and the time taken to get there. Use this to work out the speed.

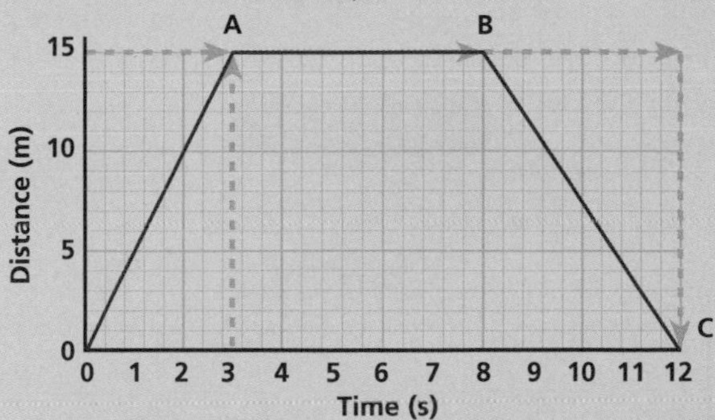


So, the object travelled at 5m/s for 3 seconds, remained stationary for 5 seconds then travelled backwards at 3.75m/s for 4 seconds until it reached the starting point.

**Distance–Time Graphs for Non-Uniform Speed**

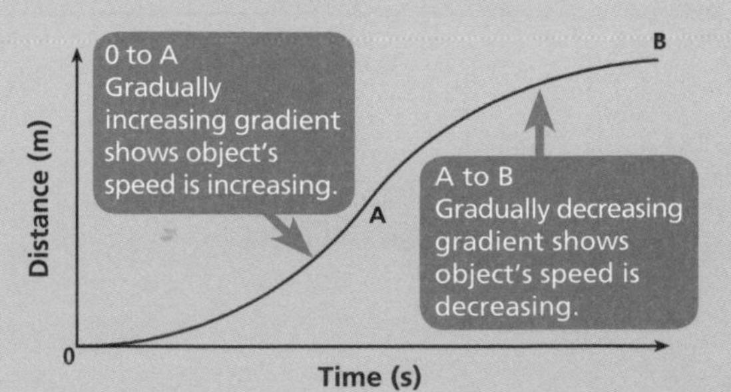

## Speed–Time Graphs

A speed–time graph plots a person's or object's speed at different times.

The slope of a **speed–time graph** represents the acceleration of an object.

A constant acceleration gradually increases the speed at a uniform rate. The steeper the slope (from bottom left to top right), the bigger the acceleration.

A negative gradient slope (from top left to bottom right) indicates **deceleration** (decreasing speed).

**1 Constant Speed**

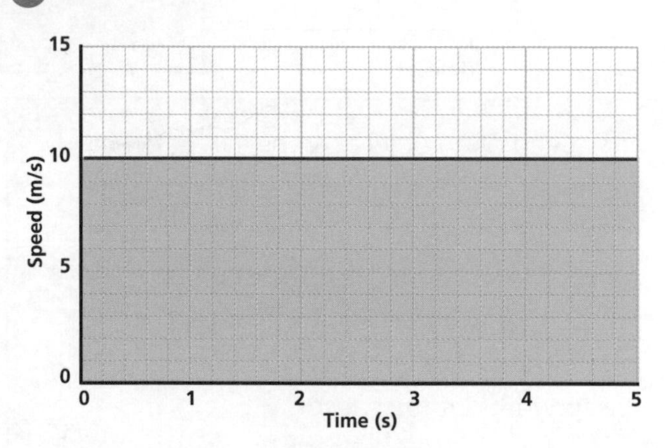


Object is moving at a constant speed of 10m/s, i.e. it is not accelerating.

$\frac{10-10}{5} = 0m/s^2$

**2 Increasing Speed**

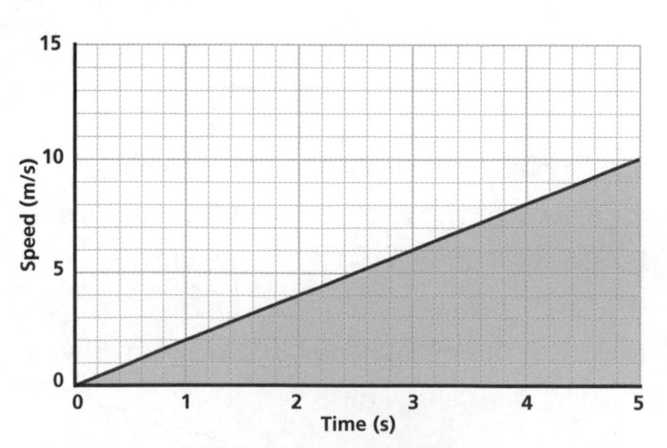


Object's constant acceleration is $2m/s^2$.

$\frac{10-0}{5} = 2m/s^2$

**3 Increasing Speed**

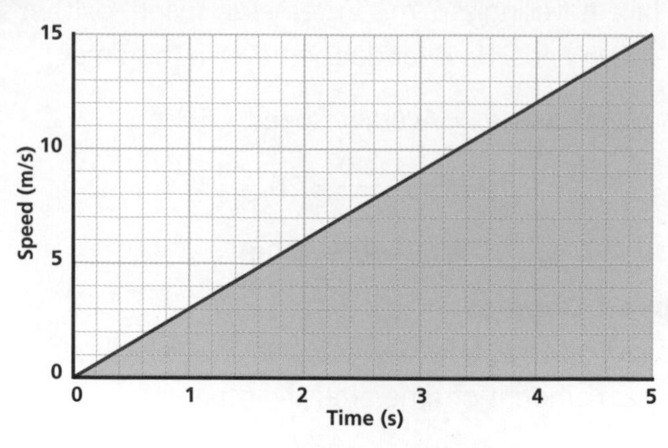


Object's constant acceleration is $3m/s^2$.

$\frac{15-0}{5} = 3m/s^2$

**4 Decreasing Speed**

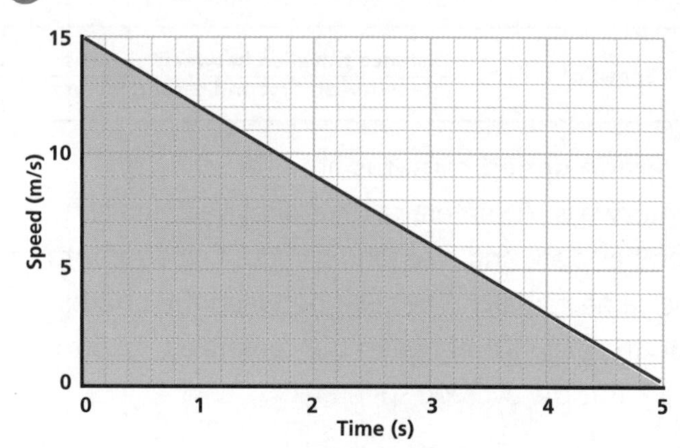


Object's constant acceleration is $-3m/s^2$.

$\frac{0-15}{5} = -3m/s^2$

HT The area underneath the line in a speed–time graph represents the total distance travelled.

For example, the area under the line in graph 3 ($\frac{1}{2} \times 15 \times 5 = 37.5m$) is greater than the area under the line in graph 2 ($\frac{1}{2} \times 10 \times 5 = 25m$).

This means that the distance travelled by the object in graph 3 is greater than the distance travelled by the object in graph 2 for the same time period.

HT The **acceleration** of an object can be calculated by working out the **gradient** of a **speed–time graph**: the steeper the gradient, the bigger the acceleration.

Take any two points on the graph and read off the change in speed over the chosen period, and the time taken for this change.

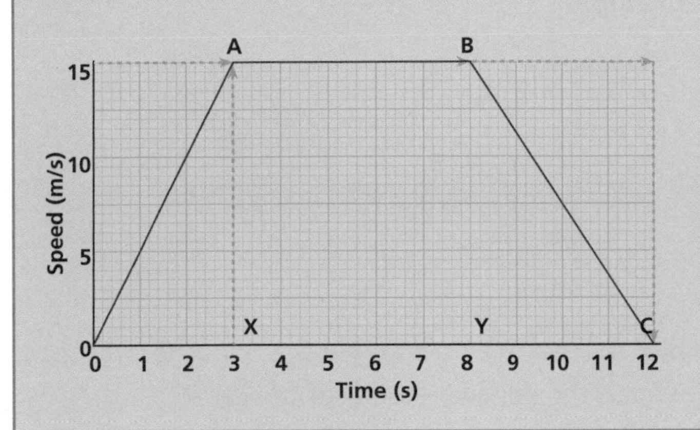


So, the object accelerated at $5m/s^2$ for 3 seconds, then travelled at a constant speed of 15m/s for 5 seconds (zero acceleration), and then decelerated at a rate of $3.75m/s^2$ for 4 seconds.

Distance travelled = Area of 0AX + Area of ABYX + Area of BCY

$= (\frac{1}{2} \times 3 \times 15) + (5 \times 15) + (\frac{1}{2} \times 4 \times 15) =$ **127.5m**

## Speed–Time Graph for Non-Uniform Motion

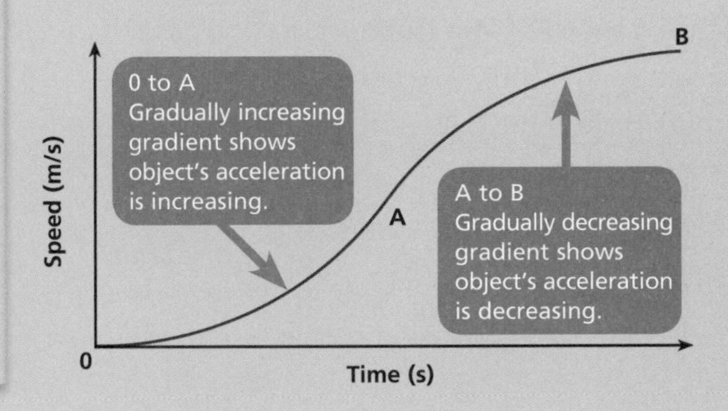


## Acceleration

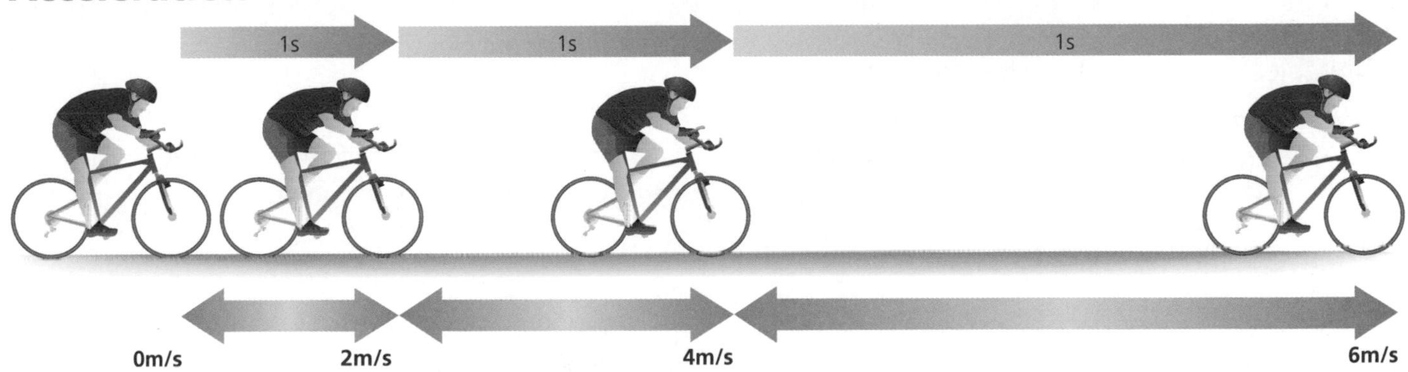


The **acceleration** of an object is the **change in speed per second**. In other words, it is a measure of how quickly an object speeds up or slows down.

**Acceleration** has the unit **metres per second per second ($m/s^2$)**. Since the cyclist in the illustration above increases his speed by 2 metres per second every second, we can say that his acceleration is $2m/s^2$ (2 metres per second, per second).

**Acceleration** is speeding up. **Deceleration** is slowing down (also called negative acceleration). A greater change in speed in a given time means a bigger acceleration.

To work out the acceleration of any moving object two things must be known:

- the change in speed
- the time taken for this change in speed.

The acceleration or deceleration of an object can be calculated using the formula:

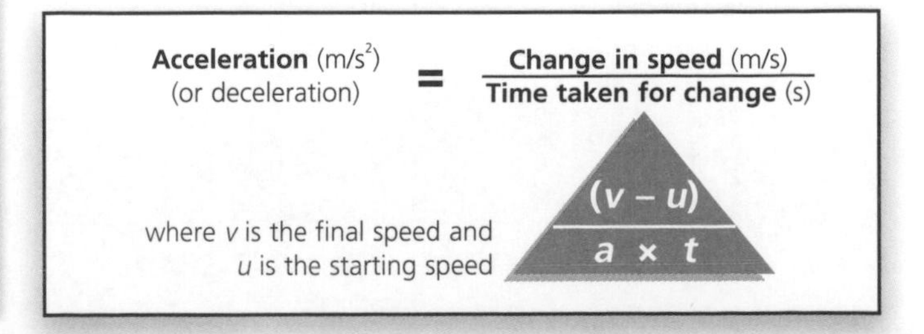

## Acceleration (cont)

There are two points to be aware of:

- The cyclist in the diagram on page 81 increases his speed by the **same amount** every second, which means the **distance** he travels each second increases.
- Deceleration is simply a negative acceleration, i.e. it describes an object that is slowing down.

### Example 1

**a)** A cyclist accelerates uniformly from rest and reaches a speed of 10m/s after 5 seconds. He then decelerates uniformly and comes to rest in a further 10 seconds. Calculate his acceleration in both cases. Use the formula:

$a = \frac{(v - u)}{t}$  $a = \frac{10 - 0\text{m/s}}{5s}$  $a = \mathbf{2m/s^2}$

**b)** Calculate his deceleration. Again, use the formula:

$a = \frac{(v - u)}{t}$  $a = \frac{0 - 10\text{m/s}}{10\text{s}}$  $a = \mathbf{-1m/s^2}$

### Example 2

An object at rest falls from the top of a building with an acceleration of 10m/s². It hits the ground with a speed of 25m/s. Calculate how long the object takes to fall.

Rearrange the formula:

$$\text{Time taken} = \frac{\text{Change in speed}}{\text{Acceleration}} = \frac{25\text{m/s} - 0}{10\text{m/s}^2} = \mathbf{2.5s}$$

### Example 3

A car accelerates at 1.5m/s² for 12 seconds. If the initial speed of the car was 10m/s, calculate the speed of the car after the acceleration.

Rearrange the formula:

Change in speed = Acceleration × Time taken
= 1.5m/s² × 12s
= **18m/s**

Speed of car after acceleration = Initial speed + Change in speed
= 10m/s + 18m/s = **28m/s**

## Relative Velocity

When two objects move in opposite directions at the same speed, their velocities have the same magnitude but opposite signs.

Relative velocity can be calculated using the formula:

**Relative velocity = Velocity 1 − Velocity 2**

### Example

**a)** A car travelling at 20m/s is following a car travelling at 15m/s. What is the relative velocity of the first car?

Relative velocity = Velocity 1 – Velocity 2
= 20 – 15
= **5m/s**

**b)** If the cars are travelling in the opposite directions, what is the relative velocity of the cars?

Relative velocity = Velocity 1 – Velocity 2
= 20 – –15
= **35m/s**

Remember:

- Negative acceleration is **deceleration**.
- Negative speed means going in the opposite direction.
- If we need to note the direction an object is moving in we use **velocity**.
- **Velocity** includes information about speed and direction – velocity is speed in a given direction.

*N.B. Acceleration can also involve a change of direction and/or speed. It is a vector quantity.*

*Acceleration is caused by a force. If the force acts in a different direction to that in which the object is moving – say from the side by the wind – then the object will change direction and experience an acceleration.*

## Forces in Action

**Forces** are **pushes** or **pulls**. They are measured in **newtons (N)** and may be different in size and act in different directions. Forces can cause objects to speed up or slow down, for example:
- **gravity** causes an apple falling from a tree to speed up as it falls; it accelerates
- **friction** causes a car to slow down; it decelerates
- **air resistance** causes a parachutist to slow down after opening a parachute; she decelerates.

When the air resistance on the parachute exactly cancels out the downward force of gravity, the parachute will stay at the same speed as it falls. It does not speed up or slow down because the forces are balanced.

The relationship between force, mass and acceleration is shown in the formula:

**Force** (N) = **Mass** (kg) × **Acceleration** ($m/s^2$)

F
m × a

From this formula, we can define a newton (N) as the force needed to give a mass of one kilogram an acceleration of one metre per second per second ($1m/s^2$).

A bigger force on a given mass will cause a bigger acceleration.

**Example**

A trolley of mass 400kg is accelerating at $0.5m/s^2$. What force is needed to achieve this acceleration?

Use the formula:

Force = Mass × Acceleration
= 400kg × $0.5m/s^2$
**= 200N**

The girl in the diagram below is standing on the ground. She is being pulled down to the ground by gravity, and the ground is pushing up with an equal force – the reaction force.

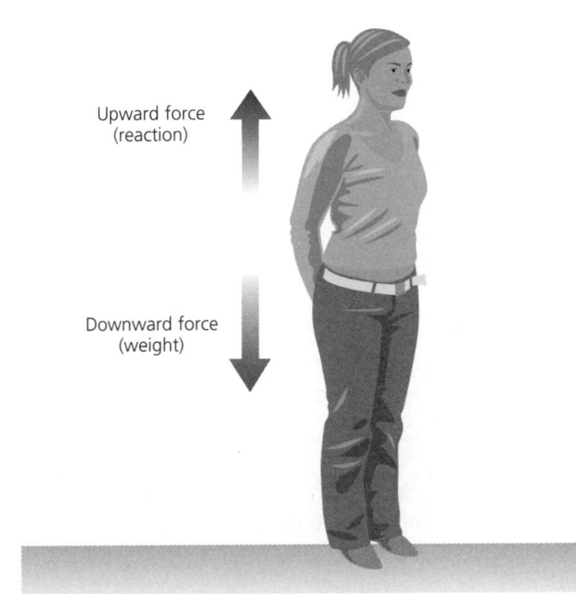


## Stopping Distance

The **stopping distance** of a vehicle depends on:
- the **thinking distance** – the distance travelled by the vehicle from the point the driver realises he needs to brake to when he applies the brakes
- the **braking distance** – the distance travelled by the vehicle from the point the driver applies the brakes to the point at which the vehicle actually stops.

**Stopping distance = Thinking distance + Braking distance**

The **thinking distance** is increased if:
- the vehicle is travelling faster
- the driver is ill, tired or under the influence of alcohol or drugs
- the driver is distracted or is not fully concentrating
- there is poor visibility – this delays the time before the driver realises he needs to apply the brakes.

## Stopping Distance (cont)

The **braking distance** is increased if:

- the vehicle is travelling faster
- there are poor weather or road conditions, e.g. if it is wet, slippery or icy
- the vehicle is in poor condition, e.g. brakes and tyres are worn out, tyres are not inflated properly.

A longer thinking distance and a longer braking distance means the car takes longer to stop (longer stopping distance).

Drivers should not drive too close to the car in front and should always keep to the speed limit (or below if road or driving conditions are poor) so that the stopping distance is long enough to avoid an accident.

The illustration below shows how the thinking distance and braking distance of a vehicle under normal driving conditions depend on the vehicle's speed.

It takes much longer to stop at faster speeds, which is why, for safety, drivers should always:

- obey the speed limits
- keep their distance from the car in front (more than the stopping distance)
- allow extra room between cars, or drive more slowly in bad weather or poor road conditions.

HT As the speed of a car increases:

- the driver's thinking distance increases linearly
- the driver's braking distance increases as the square of the speed:
  - speed doubles, braking distance increases by a factor of 4
  - speed trebles, braking distance increases by a factor of 9.

The braking distance of a vehicle is increased if:

- the **mass** of the vehicle is **increased** – a loaded vehicle, i.e. a vehicle with passengers, baggage etc. has a higher kinetic energy, which increases the braking distance
- the **friction** between the tyres and the road is **decreased** – a wet or greasy road surface reduces the amount of friction between the tyres and the road, which increases the braking distance
- the **braking force** applied is **decreased** – a smaller force exerted by the brake pads on the wheel discs increases the braking distance
- the vehicle is **travelling faster** – a faster vehicle has higher kinetic energy, which increases the braking distance.

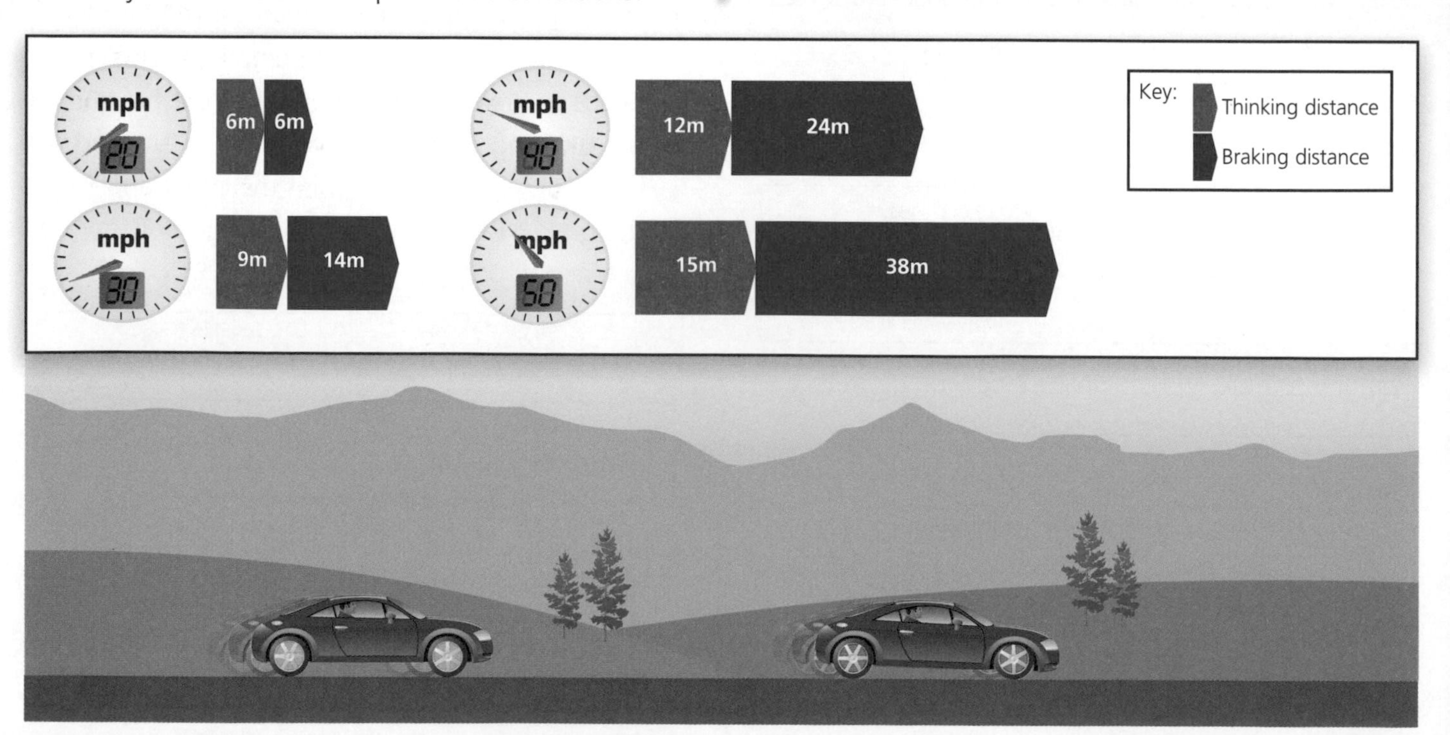

## Work

**Work** is done whenever a force moves an object. Every day you are doing work and developing **power**, for example in activities like:

- lifting weights
- climbing stairs
- pulling a wheelie bin
- pushing a shopping trolley.

Energy is needed to do work. Both energy and work are measured in **joules, J**:

**Work done** (J) **=** **Energy transferred** (J)

The amount of work done depends on:

- the **size** of the force in newtons
- the **distance** the object is moved in metres in the direction of the force.

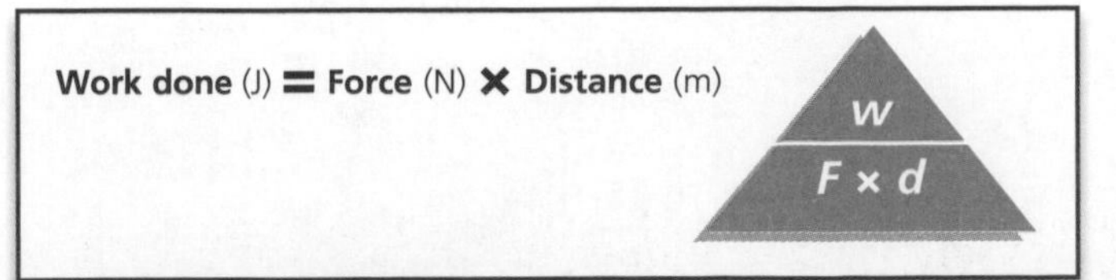


**Example**

A man pushes a car with a steady force of 250N. The car moves a distance of 20m. How much work does the man do?

Work done = Force × Distance
= 250N × 20m
**= 5000J (or 5kJ)**

## Weight

If you are lifting your own weight, you are doing work.

**Weight** (N) **=** **Mass** (kg) **×** **Gravitational field strength** (N/kg)

**Example**

A person has a mass of 60kg. Calculate their weight on Earth where the gravitational field strength is 10N/kg. Use the equation:

Weight = Mass × Gravitational field strength
= 60kg × 10N/kg
**= 600N**

## Fuel Consumption

Different cars have different power ratings (usually expressed as the number of seconds to get from 0mph to 60mph) and different engine sizes (usually expressed as litres – so a larger engine uses more fuel).

Car **fuel consumption** depends on:

- the engine size – larger engines use more fuel
- the force required to work against friction
- driving style and speed – more accelerating and braking uses more fuel
- road conditions
- tyre pressure.

Obviously, using more fuel has a bigger impact on the environment – using more resources, creating more pollutants in the exhaust gases and costing more money.

## Power

**Power** is a measure of how quickly work is done, i.e. the work done per second. The unit of power is the **watt, W**.

Some cars have much higher **power ratings** than others and they may also use far more fuel. High fuel consumption is expensive for the driver and is also damaging to the environment.

Power, work done and time taken are linked by the formula:

$$\textbf{Power}\ (W) = \frac{\textbf{Work done}\ (J)}{\textbf{Time}\ (s)}$$

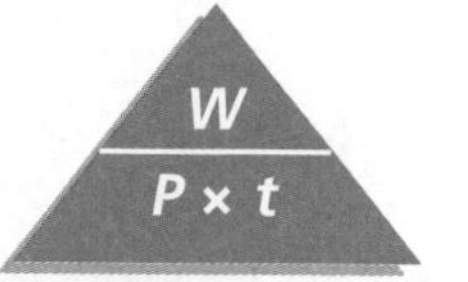


### Example 1

A girl of mass 60kg runs up a flight of stairs to a height of 4m in 8 seconds. What is her power?

Weight of girl = Mass × Gravitational field strength
$W$ = 60kg × 10N/kg
**= 600N**

Work done running up stairs:
Work done = Force × Distance
= 600N × 4m
**= 2400J**

So the girl does 2400 joules of work when she runs up a flight of stairs in 8 seconds. Calculate her power.

$$\text{Power} = \frac{\text{Work done}}{\text{Time}} = \frac{2400\text{J}}{8\text{s}}$$

**= 300W**

### Example 2

A crane does 200 000J of work when it lifts a load of 25 000N. The power of the crane is 50kW.

**a)** Calculate the distance moved by the load.
Rearrange the formula:

$$\text{Distance} = \frac{\text{Work done}}{\text{Force}} = \frac{200\ 000\text{J}}{25\ 000\text{N}}$$

**= 8m**

**b)** Calculate the time taken to move the load.
Rearrange the formula:

$$\text{Time} = \frac{\text{Work done}}{\text{Power}} = \frac{200\ 000\text{J}}{50\ 000\text{W}}$$

**= 4s**

Power can also be calculated using **Force × Speed**.

This comes from:

$$\text{Power} = \frac{\text{Work done}}{\text{Time}}, \text{ and}$$

$$\text{Work done} = \text{Force} \times \text{Distance}$$

Therefore, $\text{Power} = \text{Force} \times \frac{\text{Distance}}{\text{Time}}$

Thus, Power = Force × Speed

P
F | s

## Kinetic Energy

**Kinetic energy** is the energy an object has because of its movement, i.e. if it is moving it has got kinetic energy. Kinetic energy is measured in joules.

The following all have kinetic energy:

- a ball rolling along the ground
- a car travelling along a road
- a person running.

Kinetic energy depends on two things:

- the **mass** of the object (kg)
- the **speed** of the object (m/s).

A moving car has kinetic energy because it has both mass and speed. If the car moves with a faster speed it has more kinetic energy (providing its mass has not changed).

HT If the mass of the car is greater (e.g. there are more people inside it, or it is a larger vehicle, e.g. a truck), it may have more kinetic energy even if its speed is less than that of another car.

A car with more kinetic energy will have a longer braking distance because there is more energy to get rid of using the brakes (which is why they get hot).

You can calculate kinetic energy using the formula:

$$\textbf{Kinetic energy}\ (J) = \frac{1}{2} \times \textbf{Mass}\ (kg) \times \textbf{Speed}^2\ (m/s)^2$$

**Example 1**
A car of mass 1000kg is moving at a speed of 10m/s. How much kinetic energy does it have?

Use the formula:

$$\text{Kinetic energy} = \frac{1}{2} \times \text{Mass} \times \text{Speed}^2$$
$$= \frac{1}{2} \times 1000\text{kg} \times (10\text{m/s})^2$$
$$= \mathbf{50\,000J\ (or\ 50kJ)}$$

**Example 2**
Calculate the kinetic energy of a toy car of mass 400g moving at a speed of 0.5m/s. Use the formula:

$$\text{Kinetic energy} = \frac{1}{2} \times \text{Mass} \times \text{Speed}^2$$
$$= \frac{1}{2} \times 0.4\text{kg} \times (0.5\text{m/s})^2$$
$$= \mathbf{0.05J}$$

## Fuels

Although most cars rely on **fossil fuels** such as **petrol** or **diesel** for their energy, it is also possible to use **biodiesel**. Biodiesel is a liquid fuel made from plants. The main disadvantage is that these plants are grown in fields that could have been used to grow other plants for food.

**Electricity** can also be used. **Battery-driven** cars are already on our roads and **solar-powered** cars with solar panels on their roofs will soon follow. Solar-powered cars would produce no pollution. Unlike cars powered by fossil fuels, cars powered by electricity do not pollute at the point of use. However, their batteries are recharged using electricity that is generated in power stations, which does cause pollution.

There may be an overall reduction in carbon dioxide production to generate this energy when power stations are better set up to control pollution and emissions. However, in the UK we do not have enough power stations to produce enough electricity if everyone had electric cars and the National Grid could not transmit enough current to charge all the batteries.

## Fuels (cont)

Our supplies of fossil fuels are running low and pollution and **climate change** are now such big problems that cars will soon have to run on less polluting fuels such as biofuels, electricity or **hydrogen fuel cells**.

HT Biofuels use fuel from crops so the carbon dioxide emitted is balanced by the carbon dioxide taken in by the crops when they were growing.

However, producing biofuels and transporting them to users requires conventional fossil fuels. This means that pollution is produced in production, even if there is less pollution produced at the point of use.

Solar-powered vehicles do not pollute because they generate electricity from the Sun.

However, they are unreliable in the UK because there is so little sunlight. Producing solar panels does produce pollution and uses up valuable resources, so they are not completely pollution-free.

## Vehicle Fuel Consumption

Designing better vehicle shapes can reduce their air resistance, which improves their fuel consumption as well as giving them higher top speeds, for example:

- sports cars are wedge shaped
- lorries and caravans often have deflectors on top to given them a less 'boxy' shape.

Fuel consumption is increased by increased air resistance, such as roof boxes on top of cars and driving with the windows down.

HT Car **fuel consumption** depends on:

- the energy required to increase kinetic energy
- the energy required to work against friction
- driving style and speed – more accelerating and braking uses more fuel
- different road conditions
- tyre pressure.

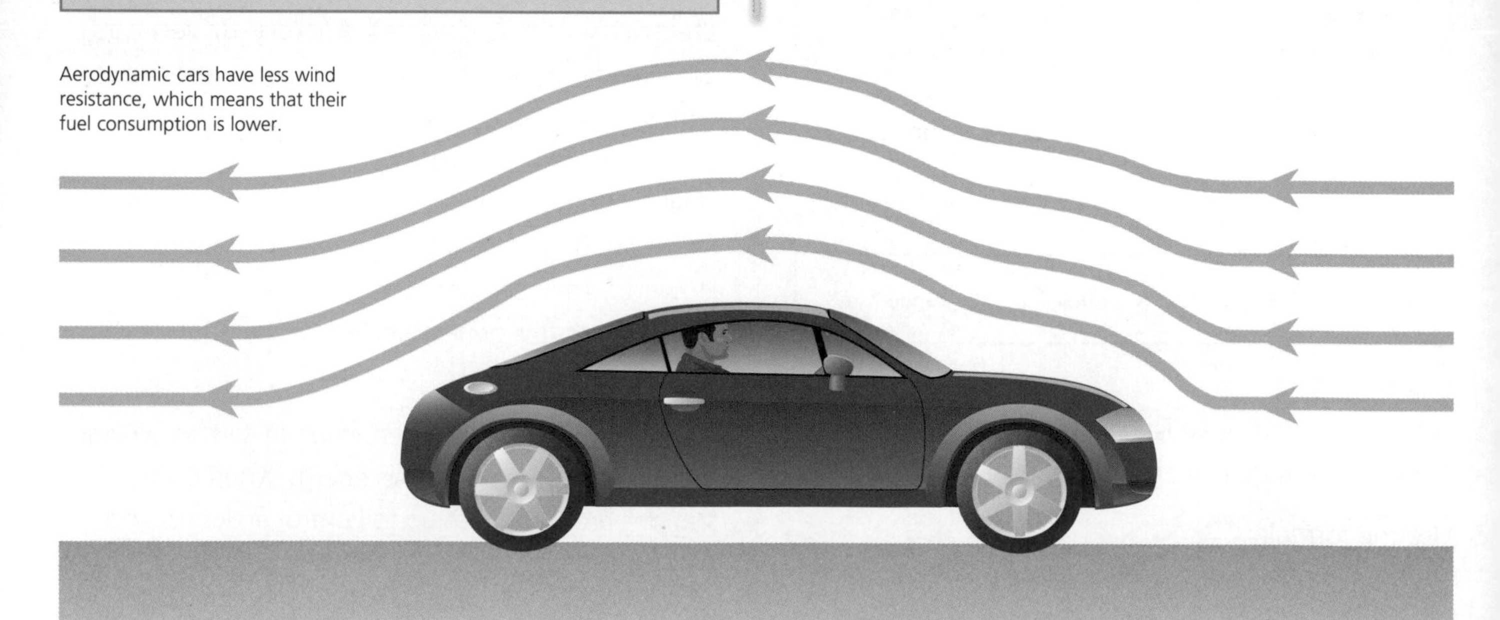

Aerodynamic cars have less wind resistance, which means that their fuel consumption is lower.

## Momentum

**Momentum** is calculated using:

$$\underset{(\text{kg m/s})}{\textbf{Momentum}} = \underset{(\text{kg})}{\textbf{Mass}} \times \underset{(\text{m/s})}{\textbf{Velocity}}$$

So if a vehicle has more mass or is travelling faster it has more momentum.

**Example 1**
Calculate the momentum of a bullet of mass 10g travelling at 300m/s.

Momentum = Mass × Velocity
= 0.01kg × 300m/s = **3 kg m/s**

**Example 2**
If a ball of mass 300g has a momentum the same as the bullet above, how fast is it travelling?

$$\text{Velocity} = \frac{\text{Momentum}}{\text{Mass}}$$

$$\text{Velocity} = \frac{3\text{kg m/s}}{0.3\text{kg}} = \mathbf{10\ m/s}$$

## Force

**Force** can be calculated using:

$$\textbf{Force}\ (\text{N}) = \frac{\textbf{Change of momentum}\ (\text{kg m/s})}{\textbf{Time}\ (\text{s})}$$

**Example**
In example 1 above, if the bullet is shot from the gun and accelerates to its full speed in 0.01 seconds then what is the force on it? The bullet's momentum goes from 0 to 3 kg m/s in 0.01 seconds.

$$\text{Force} = \frac{\text{Momentum change}}{\text{Time}}$$

$$= \frac{3-0}{0.01} = \mathbf{300N}$$

## Car Safety Features

In a car accident the driver and passengers decelerate very quickly and therefore experience a large change of momentum as their velocity decreases to zero rapidly. This means they experience a very large force, which can cause injuries.

Modern cars have **safety features** that **absorb energy** during a collision, thereby reducing the injuries. These safety features include:

- **seat belts** – that stretch slightly and also have a mechanism in the base which means the person's energy is absorbed
- **air bags** – to cushion the impact for the driver and front passenger
- **brakes** – to reduce the speed of the car by transferring kinetic energy to heat energy
- **crumple zone** – a region of the car designed specifically to 'crumple' during a collision. This absorbs a lot of the energy in a crash, reducing the danger to the people in the car.

Crumple zones, seat belts and air bags all **change shape** on impact to **absorb energy** and therefore reduce the risk of injury to the people in the car. They also lengthen the time that the passenger takes to stop. This means that a passenger's momentum changes over a longer time so they experience less force and suffer fewer injuries.

Seat belts have to be replaced after a crash because they can be damaged (perhaps stretched) by the force of a body against them. The benefit of seat belts is the potential to reduce injury. There is a very slight risk that a passenger might be trapped in a car by their seat belt after an accident and not be able to get out. On balance though, it is always better to wear a seat belt.

HT Newton's second law of motion, often written as $F = ma$ can explain why force is calculated as change of momentum divided by time.

Force = Mass × Acceleration, and

$$\text{Acceleration} = \frac{\text{Change in velocity}}{\text{Time}}$$

$$\text{So, Force} = \text{Mass} \times \frac{\text{Change in velocity}}{\text{Time}}$$

$$\text{Or, Force} = \frac{\text{Mass} \times \text{Change in velocity}}{\text{Time}}$$

But, Mass × Velocity = Momentum

$$\text{So, Force} = \frac{\text{Change in momentum}}{\text{Time}}$$

## Safety Features to Prevent Accidents

Some safety features on a car are there to **prevent accidents** (passive safety features). For example:

- **Anti-lock braking systems** (ABS) – prevent the tyres from skidding, which means the vehicle stops more quickly and allows the driver to remain in control of the steering.
- **Traction control** – prevents the car from skidding when accelerating, so the driver can quickly escape from a dangerous situation.
- **Electric windows** – make it easier for the driver to open or close the windows when driving, causing less of a distraction.
- **Paddle shift controls** – allow the driver to keep both hands on the steering wheel when changing gear and adjusting the radio.

## Safety Features to Protect Passengers

Some safety features **protect the car's occupants** (active safety features). For example:

- **Crumple zones** – the car bonnet crumples absorbing energy making the passengers slow down more gently.
- **Collapsible steering columns** – so that the driver isn't injured when he/she hits the steering wheel in an accident.
- **Airbags** – protect passengers by inflating to provide a cushioning effect to absorb energy instead of the passengers hitting the dashboard.
- **Seat belts** – hold the passengers in place so that they are not flung through the windscreen on impact.
- **Safety cage** – a metal cage which strengthens the cabin section of the car to prevent the vehicle from collapsing when it is upside down or rolling.

Safety Cage Reinforces Body of Car

HT Wearing a seat belt is compulsory in the UK in both the front and back seats (if seat belts are fitted). This law was brought in to protect people from injuries and to reduce the cost to the National Health Service of treating people after car accidents.

Data from crash tests where dummy drivers and passengers are observed during a controlled crash can identify how the worst injuries happen. This allows car manufacturers to include improved safety features such as side impact bars.

## HT Reducing Stopping Forces

The stopping **forces** experienced by the people in the car in a collision can be reduced by:

- increasing the stopping or collision time
- increasing the stopping or collision distance
- decreasing acceleration.

Because force is calculated by **change of momentum divided by time**, this means that the force on a passenger's body is smaller and so the passenger is less likely to be injured badly.

All the safety features mentioned on this page perform one or more of the above tasks. By reducing the stopping forces on the people in the car, they reduce the risk of injury.

**Anti-lock braking systems** prevent the tyres from skidding during heavy braking by pumping on and off rather than holding the wheel tight. This increases the area of the tyres in contact with the road and gives the driver better control of the car. This often reduces the overall braking distance.

## Friction and Air Resistance

**Frictional forces**, such as **drag**, **friction** and **air resistance**, can act against the movement of an object, slowing it down. They reduce the object's kinetic energy. This effect can be reduced by:

- changing the **shape** (to increase or decrease air resistance)
- using a **lubricant** (to make the object slide with less resistance).

The **shape** of an object can influence its top speed:

- **Shuttlecocks** in badminton are designed to increase air resistance so they travel slower.
- **Parachutes** are designed to have a larger surface area to increase air resistance.
- **Flying squirrels** have skin between their legs and arms to create a large surface area to stop them falling too quickly between tree branches.

Falling objects speed up as they fall because they are pulled towards the centre of the Earth by **gravity**.

## Terminal Speed

Falling objects experience two forces:

- the downward force of **weight**, *W*, which always stays the same on Earth and happens because of the pull of the Earth's gravity
- the upward force of **air resistance**, *R*, or drag, which increases the faster the object falls.

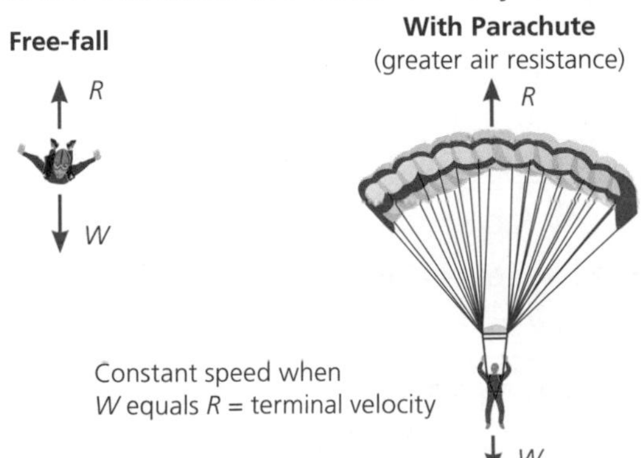


When the weight is bigger than the air resistance then the object speeds up. When the air resistance is bigger than the weight, then the object slows down. If the air resistance is the same as the weight then the forces are balanced and the object continues at the same speed – **its terminal velocity.**

## Weight and Gravity

Objects **fall** because of their **weight** and they get faster as they fall. Weight is a **force** pulling towards the centre of a planet – in our case, Earth. The strength of this force depends on the **gravity** of the planet. On the Earth the acceleration caused by gravity is the same on all objects at any particular point on the Earth's surface.

If there is no atmosphere there can be no air resistance (drag). This means that falling objects can not slow down. This happens on the Moon; all objects fall at the same rate on the Moon (nothing can float!).

HT The **Earth's gravitational field strength** (or acceleration due to gravity) is unaffected by atmospheric changes but can be slightly different at different points on the Earth's surface. For example, it will be different at the top of a mountain compared with at the bottom of a mine. But the Earth's field exerts the force required to accelerate the object at exactly the same rate ($9.8 m/s^2$).

HT Drag on an object varies. There is more drag on an object with a larger surface area and on a faster object.

When an object's drag force equals its driving force then the two will balance and the object travels at a constant speed, called the **terminal velocity**.

When a skydiver jumps out of an aeroplane, the speed of his descent can be considered:

- before his parachute opens (i.e. free-fall)
- after his parachute opens (air resistance is increased due to the surface area of his parachute).

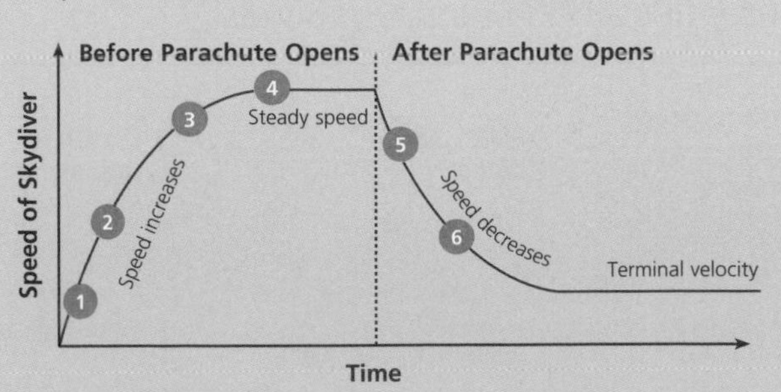

## Gravitational Potential Energy

**Gravitational potential energy** (GPE) is the energy that an object has due to its position in the Earth's gravitational field. If an object can fall (e.g. a diver standing on a diving board before jumping off) it has gravitational potential energy.

A person standing on a higher diving board will have more gravitational potential energy than another person standing on a lower diving board (providing they have the same mass). This is because he is further away from the ground (see 1).

A heavier person standing on the same diving board as a lighter person will have more gravitational potential energy. This is because he has a bigger mass (see 2).

When an object falls it transfers gravitational potential energy into kinetic energy (KE). This is what happens when:

- a diver jumps off the diving board
- a ball rolls down a hill
- a skydiver jumps out of a plane.

An object's KE increases (it gets faster) as its GPE reduces (it gets closer to the ground).

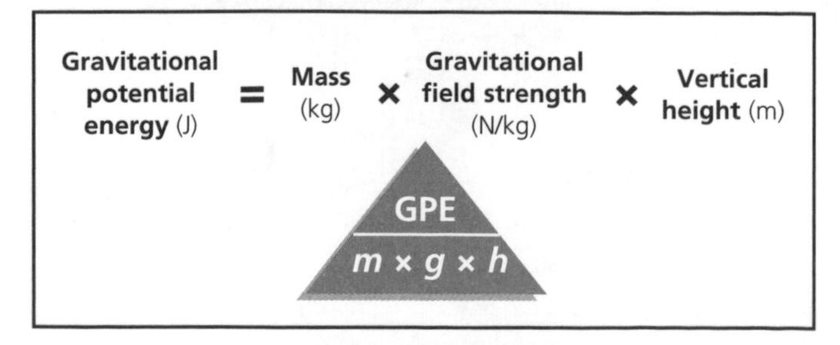


HT Gravitational field strength, $g$, is a constant. On Earth it has a value of **10N/kg**. This means that every 1kg of matter near the surface of the Earth experiences a downwards force of 10N due to gravity.

On planets where the gravitational field strength is higher, the gravitational potential energy is greater.

**Example 1**

A skier of mass 80kg gets on a ski lift that takes her from a height of 1000m to a height of 3000m above ground. By how much does her gravitational potential energy increase? Use the formula:

GPE = $m \times g \times h$

= 80kg × 10N/kg × (3000m – 1000m)

= 80kg × 10N/kg × 2000m

**= 1 600 000J (or 1600kJ)**

*N.B. Work has been done by the ski-lift motor so that the skier can increase her gravitational potential energy. In other words, work done by the motor has been transferred into her gravitational potential energy. She will transfer this gravitational potential energy into kinetic energy as she skis down the slope.*

**Example 2**

A ball is kicked vertically upwards from the ground. Its mass is 0.2kg and it increases its gravitational potential energy by 30J when it reaches the top point in its flight. What height does the ball reach? Rearrange the formula:

$$\text{Vertical height} = \frac{\text{GPE}}{\text{Mass} \times \text{Gravitational field strength}}$$

$$= \frac{30\text{J}}{0.2\text{kg} \times 10\text{N/kg}}$$

**= 15m**

## Gravitational Potential Energy and Kinetic Energy

When an object falls, its **gravitational potential energy** is transferred to **kinetic energy**. There are many theme park rides that use this transfer of energy.

1. On most roller-coasters, the cars start high up with a lot of gravitational potential energy.
2. As the cars drop, the gravitational potential energy is gradually transferred into kinetic energy.
3. The car reaches its highest speed, maximum kinetic energy, at the bottom of the slope.
4. As the car climbs the slope on the other side, kinetic energy is converted back into gravitational potential energy, and it slows down.

If the **mass** of the car is **doubled**, the **kinetic energy** also **doubles**.

If the **speed** of the car is **doubled**, the **kinetic energy quadruples**.

Increasing the **gravitational field strength**, will increase the **gravitational potential energy**, but this would require you to move the roller-coaster to a different planet!

HT When an object is falling, the GPE is transferred to KE if it speeds up.

So $mgh = \frac{1}{2}mv^2$

You can work out what height of roller-coaster you need to allow the car to achieve a certain speed.

**Example**

If you want the car to get up to a speed of 20 m/s at the end of a drop, how high must it start? Assuming that no other work is done:

$$mgh = \frac{1}{2}mv^2$$

$$h = \frac{v^2}{2g}$$

$$h = \frac{20 \times 20}{2 \times 10}$$

$$h = \frac{400}{20} = \mathbf{20m}$$

It is interesting to notice that the mass of the car and occupants makes no difference in this calculation.

When an object is falling at its terminal speed, the speed is not changing, so the kinetic energy does not increase. However, the gravitational potential energy decreases as the object does work against friction or air resistance (gravitational potential energy is transferred into heat and sound energy).

# P3 Exam Practice Questions

1. Graham makes a journey from home in his car to pick up his friend for lunch. The graph is a distance–time graph of his movements.

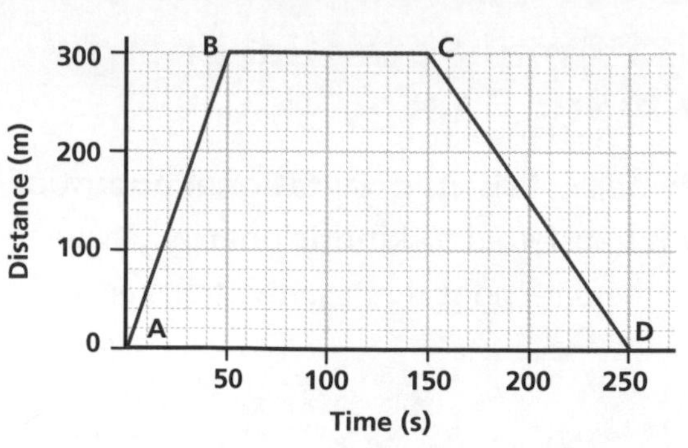


   a) Describe Graham's movements at each of the following points: **[3]**

      i) A to B  ii) B to C  iii) C to D

   b) What can you say about his speed on the way there compared to the way back? **[1]**

   c) Calculate Graham's speed during the different parts of the journey. **[2]**

   d) Draw a speed–time graph of his journey. **[3]**

2. The stopping distance of a car is the combination of the thinking distance and the braking distance.

   a) Calculate the stopping distance if the thinking distance is 35m and the braking distance is 45m. **[1]**

   b) Describe two factors for both thinking and stopping distance which would increase their size. **[4]**

   c) If a driver doesn't consider the thinking and braking distance when driving, he/she may be involved in a crash. Most cars are fitted with technology to reduce the effects of a crash and prevent injury, such as crumple zones. Explain how crumple zones work. **[3]**

3. a) Michelle has just picked up her new car and is driving it home. The mass of her car is 1300kg. She is driving at 50mph (22.3m/s) and enters a 30mph zone, where she slows to 30mph (13.4m/s). Calculate the change in momentum of the car. **[2]**

HT

   b) Whilst driving at 30mph (13.4m/s), Michelle sees a cat run out into the road in front of her and brakes suddenly. The car comes to a stop in 2.2s.

      i) Calculate the change in momentum of the car. **[2]**

      ii) What force must the brakes apply to stop the car in this time? **[3]**

4. Eva is on a roller-coaster, as shown in the diagram.

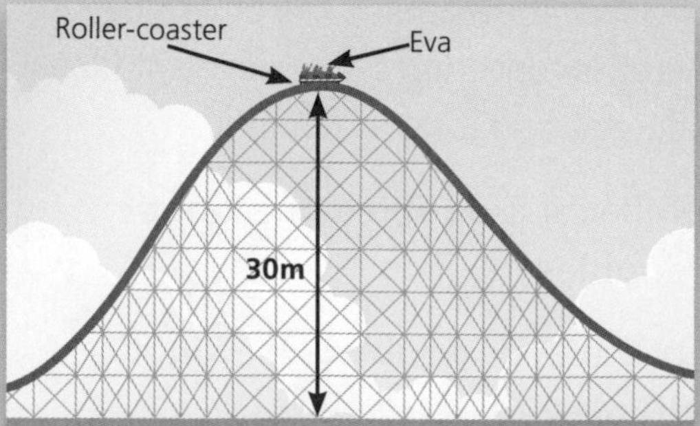

Roller-coaster
Eva
30m

   a) Calculate Eva's gravitational potential energy at the top of the peak. She weighs 80kg. Use gravity as 10N/kg. **[2]**

   b) Eva travels down the slope; all of her gravitational potential energy is converted to kinetic energy. Calculate her speed at the bottom of the slope. **[2]**

   c) After the slope, the roller-coaster stops in 2 seconds. Calculate Eva's deceleration after the slope. **[2]**

## P4: Radiation for Life

This module looks at:
- Medical physics and the importance and problems of electrostatics.
- The uses of electrostatics in medicine and in everyday life.
- Electricity and safety, including earthing and fuses.
- Ultrasound and its medical uses.
- The properties and uses of nuclear radiation.
- The uses of radioisotopes in smoke alarms, cancer treatment, tracers and radioactive dating.
- The properties of waves, including ultrasound.
- Nuclear fission and nuclear fusion, and the use of nuclear fission in producing electricity.

## Generating Static Electricity

An **insulating material** can become **electrically charged** if it is rubbed with another insulating material. The charge is called **static electricity** – the electricity stays on the material and does not move. This is due to the **transfer of electrons**.

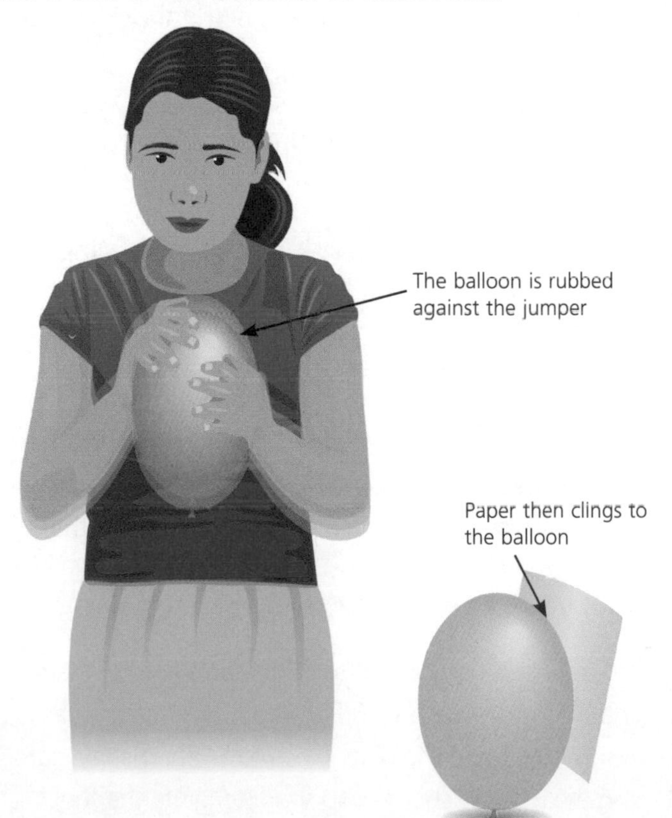


**Electrons** carry **negative charge**. The electrons transfer from one material to the other, leaving one material with a **positive** charge and the other with a **negative** charge.

You can generate static electricity by rubbing a balloon against a jumper. The electrically charged balloon will then attract light objects, e.g. pieces of paper or cork. The same effect can be obtained using a rubbed comb or strip of plastic. Dusting brushes use this effect. The brushes get charged so that they attract dust when they pass over it, making dusting more effective. Synthetic clothing can become charged up by friction between the clothing and the person's body as they move and this makes the clothing cling. When the clothing is later removed from the body, static sparks are sometimes produced.

## Discharging Static Electricity

A charged object can be **discharged** (i.e. have any charge on it removed) by **earthing** it. When an object discharges, electrons are transferred from the charged object to Earth. If you become electrically charged, earthing can result in you getting an **electrostatic shock.**

### Touching Water Pipes

A person can become charged up by friction between the soles of their feet and the floor if they are walking on an insulator such as carpet or vinyl. If they then touch a water pipe, e.g. a radiator, the charge is earthed. Discharge occurs, resulting in an electrostatic shock.

## Problems of Static Electricity

Some places, such as flour mills and petrochemical factories, have atmospheres that contain extremely flammable gases or high concentrations of fine particles. A discharge of static electricity (i.e. a spark) in these situations can lead to an **explosion**, so factories take precautions to ensure that no spark is made that could ignite the gases.

## Problems of Static Electricity (cont)

Static electricity is also dangerous in any situation where large quantities of charge could flow through your body to Earth – lightning, for example. In other situations, static electricity is not dangerous but can be a nuisance, for example:

- it can cause dirt and dust to be attracted to insulators such as television screens and computer monitors
- it can cause some fabrics to cling to your skin.

## Repulsion and Attraction

Two insulating materials with the **same charge** will **repel** each other, e.g. two Perspex rods.

For example, if a positively charged rod is held near to a suspended rod which is also positively charged, the suspended rod is repelled. The same effect would happen if both rods had a negative charge.

Two insulating materials with **different charges** will **attract** each other, e.g. a Perspex and an ebonite rod. If a negatively charged rod is held near to a suspended rod which is positively charged, the suspended rod is attracted to it. The same effect would happen if the charges were the other way round.

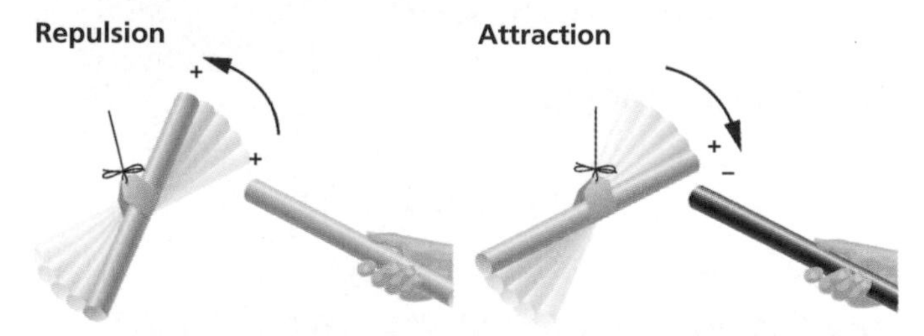


## Charging Up Objects

Electric charge (static) builds up when **electrons** (which have a **negative charge**) are rubbed off one material on to another. The material that **receives** the electrons becomes **negatively** charged due to an excess of electrons, while the material **giving up** the electrons becomes **positively** charged due to a loss of electrons.

The atoms or molecules that become charged are then called **ions**.

**Example 1**

If you rub a Perspex rod with a cloth, the Perspex gives up electrons and becomes positively charged. The cloth receives the electrons and becomes negatively charged.

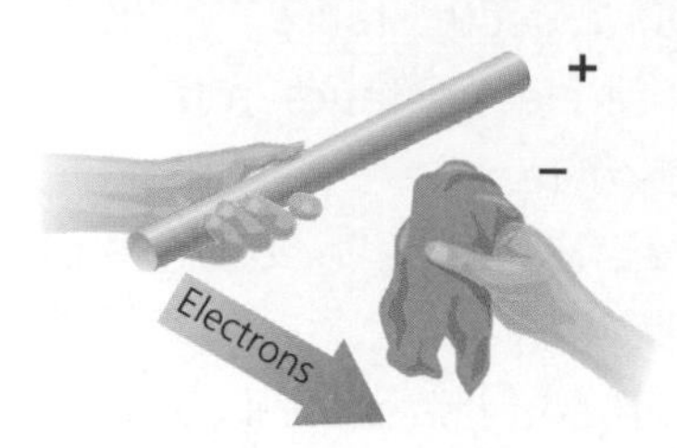


**Example 2**

If you rub an ebonite rod with fur, the fur gives up some electrons and becomes positively charged. The ebonite receives the electrons and becomes negatively charged.

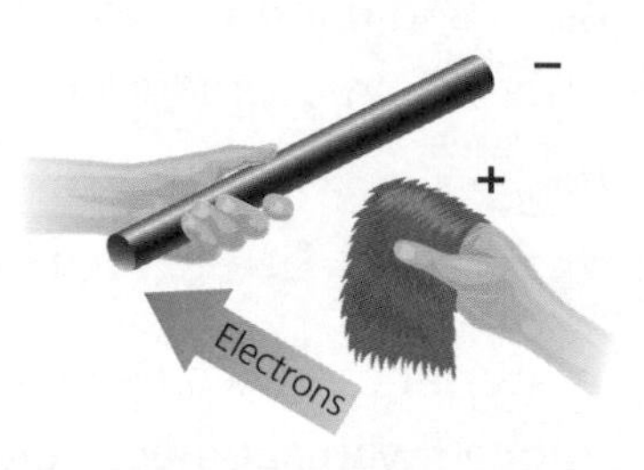


## Reducing the Danger

The chance of receiving an electric shock can be reduced by:

- ensuring that appliances are correctly earthed
- using insulation mats effectively
- wearing shoes with insulating soles.

Lorries that contain flammable gases, liquids or powders need to be earthed before unloading, as friction can cause a build-up of charge. This charge could lead to a spark, which could then ignite the flammable substance.

Anti-static sprays, liquids and cloths help to reduce the problems of static electricity by preventing the transfer of charge from one insulator to another. With no build-up of charge, there can be no discharge.

Connecting an aircraft to a fuel tanker makes sure that any charge built up on the fuel by friction, as it flows through the pipe, flows to Earth, making sure there is no discharge and spark to ignite the fuel.

## Using Static in Everyday Life

### Defibrillators

Static electricity can be used to start the heart when it has stopped:

- Two paddles are charged and are put in good electrical contact with the patient's chest.
- The defibrillator discharges into the patient.
- Charge is then passed through the patient to make the heart contract.

However, care must be taken not to shock the operator.

**Defibrillation**

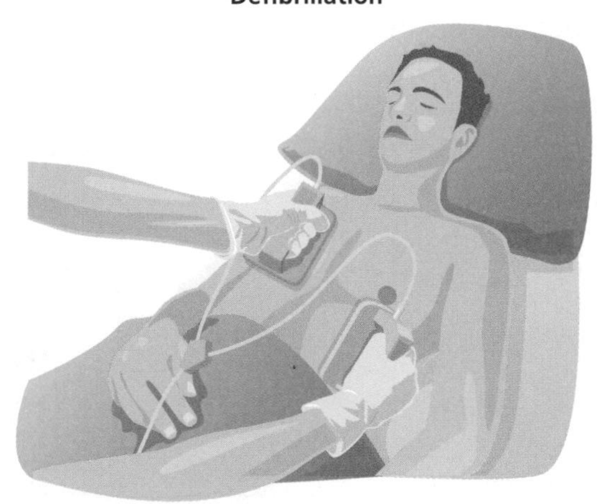

### Reducing Smoke Particles in Chimneys

Static electricity is used in electrostatic **dust precipitators** to remove smoke particles from chimneys:

- Metal plates and grids are installed in the chimney and are connected to a high voltage (potential difference). Dust passes through the grid and is charged.
- The dust particles are attracted to earthed plates, where they form larger particles. Large particles fall back down the chimney when they are heavy enough or when the plate is struck.

HT The metal plates in **dust precipitators** are at high voltage. There are also high voltage grids lower down the chimney that make the dust particles charged (they gain or lose electrons). The charged dust particles are attracted to earthed metal plates where they clump together.

**Smoke Precipitator**

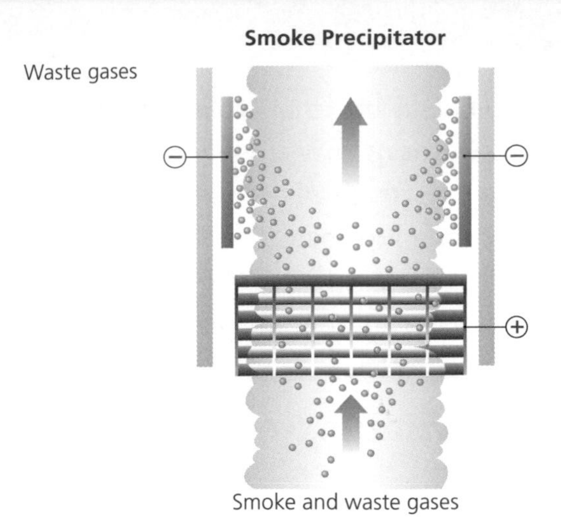


### Spraying

Electrostatics can be useful when spray painting or crop spraying. For example, when spray painting a car:

- The paint gun is negatively charged.
- Paint particles become negatively charged so they repel and spread out evenly.
- The panel that is to be sprayed is made positively charged.
- Paint from the negatively charged paint gun is attracted to the positively charged panel.

HT The negatively charged paint has gained electrons from the gun. The car panel is made positively charged by removing electrons from it to make it attract the paint. This charging process causes the paint to form a fine spray so it is applied evenly. The result is a neutral, painted car panel. It also means that less paint is wasted and even the back and sides of the object that would be in the shadow of the spray receive a coat of paint.

**Spray Painting**

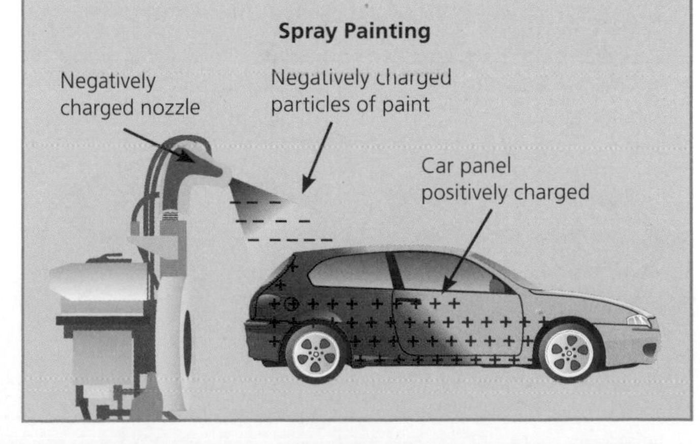

## Circuits

A **circuit** is a complete loop that allows an electric current to flow. Electrons flow around the circuit from the negative electrode of the power source to the positive electrode. However, this was only discovered relatively recently so circuit diagrams show the current flowing from the positive electrode to the negative electrode.

## Fixed and Variable Resistors

**Resistance** is a measure of how hard it is to get a **current** through a component in a circuit at a particular **voltage** (potential difference). It is measured in **ohms** (Ω).

The current through a circuit can be controlled by varying the resistance in the circuit. This can be done by using any of the following:

- A **fixed resistor** – a component whose resistance is constant. A higher resistance will give a smaller current for a particular voltage.
  A **longer** piece of resistance wire will give less current for a particular voltage.
  A **thinner** piece of wire will give less current for a particular voltage.
- A **variable resistor** (also known as a **rheostat**) – this component has a resistance that can be altered. A current that flows can be changed by simply moving the sliding contact of the variable resistor from one end to the other, as shown below.

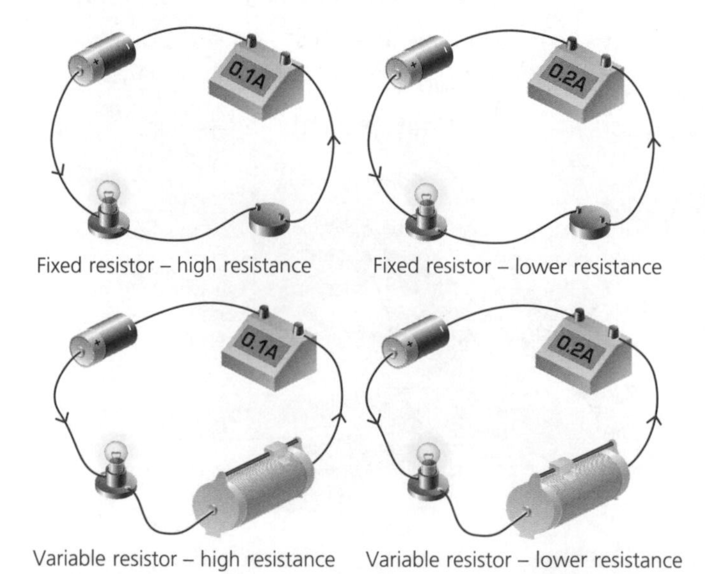


Fixed resistor – high resistance | Fixed resistor – lower resistance

Variable resistor – high resistance | Variable resistor – lower resistance

## Current, Voltage and Resistance

Current, voltage and resistance are related by the formula:

$$\text{Resistance } (\Omega) = \frac{\text{Voltage (V)}}{\text{Current (A)}}$$

where $I$ is current

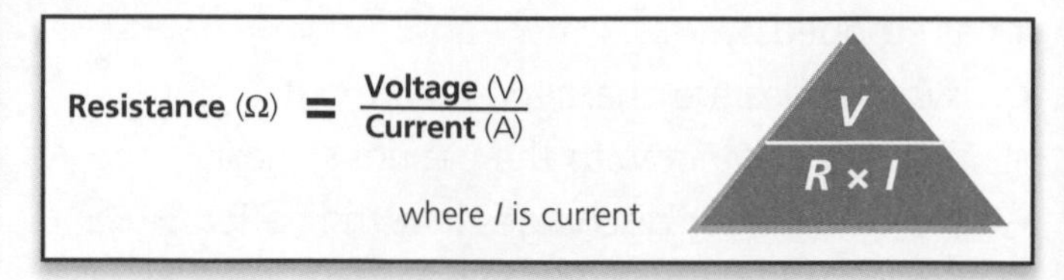


For a **given resistor, current increases as voltage increases** (and vice versa).

For a **fixed voltage, current decreases as resistance increases** (and vice versa).

**Example 1**

Calculate the resistance of the lamp in the following circuit. Use the formula:

$$\text{Resistance} = \frac{\text{Voltage}}{\text{Current}}$$

$$= \frac{3\text{V}}{0.2\text{A}}$$

$$= \mathbf{15\Omega}$$

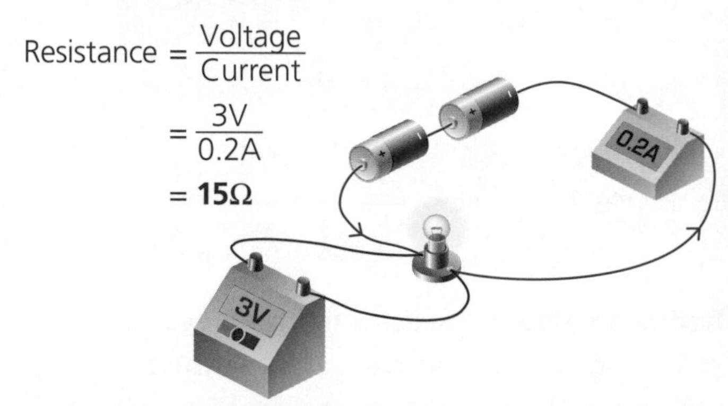


As well as being able to recall the formula above, you should be able to rearrange it to calculate **potential difference** or **current**.

**Example 2**

Calculate the reading on the ammeter in the circuit below if the bulb has a resistance of 20 ohms.

Rearrange the formula:

$$\text{Current} = \frac{\text{Voltage}}{\text{Resistance}}$$

$$= \frac{6\text{V}}{20\Omega}$$

$$= \mathbf{0.3A}$$

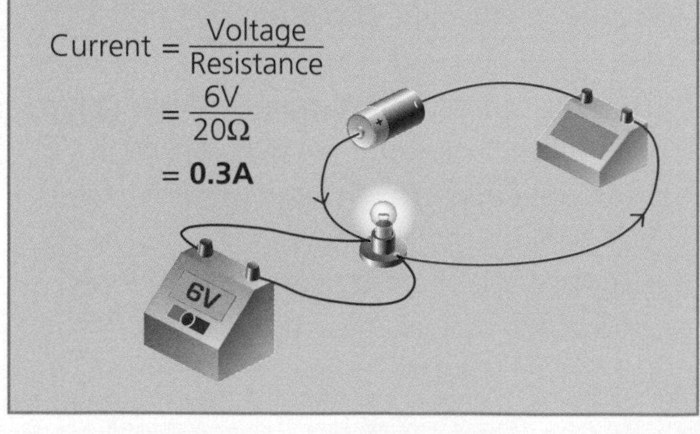

## Live, Neutral and Earth Wires

Most electrical appliances are connected to the mains electricity supply using a cable and 3-pin plug, which is inserted into a socket on the ring main circuit. Most cables contain three wires:

- The **live wire** (**brown**) carries current to the appliance at a high voltage – about 230V. (Fuses, circuit breakers and switches are always part of the live wire circuit.)
- The **neutral wire** (**blue**) completes the circuit and carries current away from the appliance.
- The **earth wire** (**green and yellow**) is the safety wire that stops the appliance becoming live.

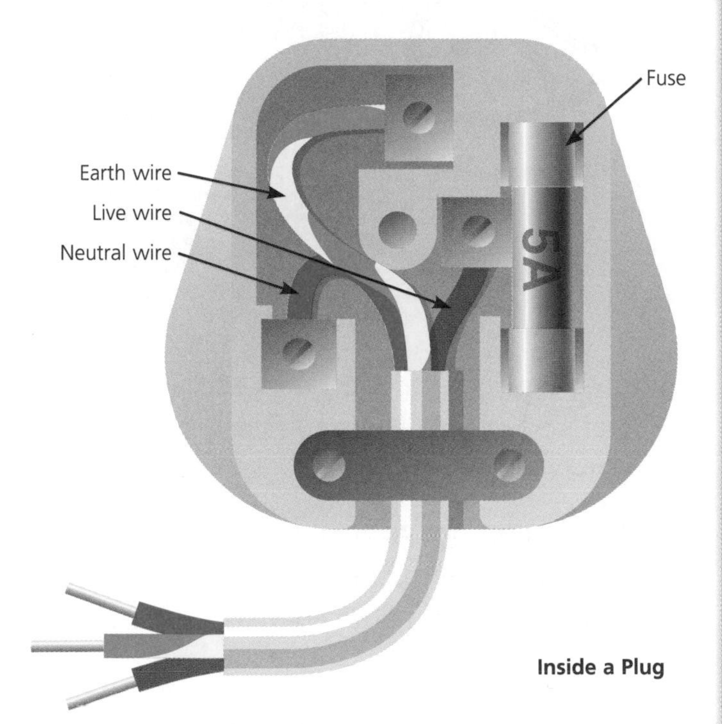

**Inside a Plug**

## Fuses and Circuit Breakers

**Fuses** and **circuit breakers** (devices that act like a fuse, but can be reset) are both safety devices. They can prevent fires, injury and death in the home by breaking the circuit of an appliance if a fault occurs.

A circuit breaker can be reset. A fuse has to be replaced.

## How a Fuse Works

A fuse is used to prevent cables or appliances from overheating and/or catching fire.

If a fault causes the current in the appliance to exceed the current rating of the fuse:

- the fuse wire gets hot and melts or breaks
- the circuit is broken so no current can flow
- the fuse prevents the flex/cable overheating
- the appliance is protected.

HT However, for this safety system to work properly, the **current rating** of the fuse must be **just above** the normal current that flows through the appliance.

Insulators wear away and wires touch

As the current increases the fuse gets hotter

The fuse melts and breaks the circuit

## Choosing the Right Fuse

The following equation is used to choose the correct fuse:

$$\text{Power (W)} = \text{Voltage (V)} \times \text{Current (A)}$$

$$\frac{P}{V \times I}$$

For example, a computer might have a power rating of 500W. The voltage used in the UK is 230V. Rearranging the formula gives:

$$\text{Current} = \frac{\text{Power}}{\text{Voltage}}$$
$$= \frac{500\text{W}}{230\text{V}}$$
$$= \mathbf{2.2A}$$

So a fuse with a rating just above 2.2A has to be used – a 3 amp fuse would be good.

HT

### Earthing

All electrical appliances with outer metal cases must have an earth wire to protect the appliance and the user.

This means that the outer case of the appliance is connected to the earth pin in the plug through the earth wire.

This is how it works:

1. A fault in the appliance causes the casing to become live, usually because the live wire touches it.
2. The circuit short-circuits (i.e. the path of the flow of charge changes) because the earth wire offers less resistance. The charge on the metal casing flows along the earth wire as a very large current.
3. The fuse wire melts (or the circuit breaker trips).
4. The circuit is broken.
5. The appliance and the user are protected.

## Double Insulation

An earthed conductor cannot become live. This is because earthing means connecting the conductor to Earth, which takes away any charge that could electrocute people.

All appliances with outer metal cases have to be earthed – this means that they have an earth wire.

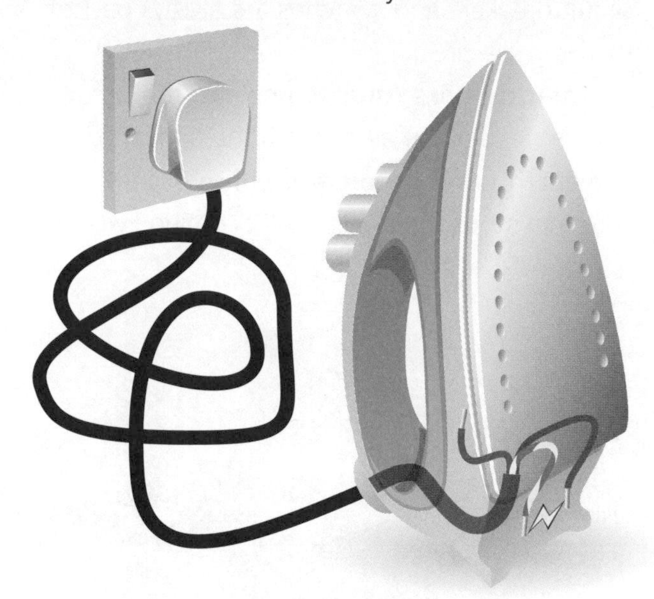

However, the outer cases of some appliances are made of insulators so their cables only need two wires; the earth wire is missing.

These appliances do not need to be earthed because they are **double insulated**; even if the live parts touch the case it does not matter because the case is an insulator.

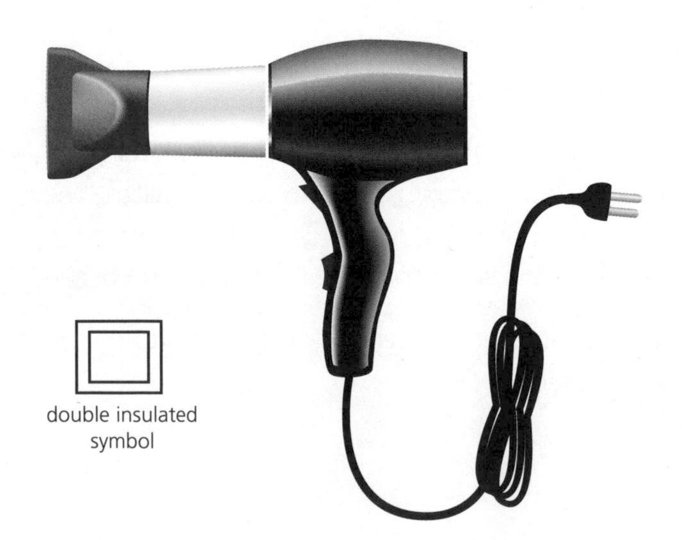

## Ultrasound

**Ultrasound** is the name given to sound waves that have frequencies higher than 20 000 hertz (Hz), i.e. above the upper threshold of the human hearing range.

Like all sound waves, ultrasound travels in a **longitudinal** wave. Longitudinal waves can be demonstrated using a slinky spring (see diagram below).

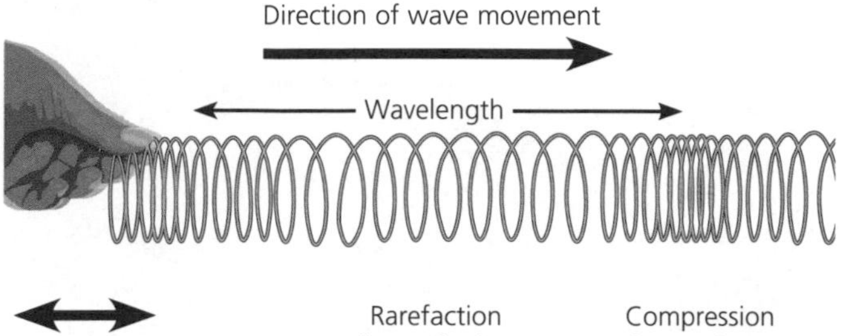


The key features of waves are:

- **amplitude** – the maximum disturbance caused by a wave
- **wavelength** – the distance between corresponding points on two successive disturbances, e.g. two compressions
- **frequency** – the number of waves produced (or that pass a particular point) in 1 second.

## Applications of Ultrasound

### Ultrasound Imaging

Ultrasound can be used in medicine to look inside people, e.g. to scan bodies, measure the speed of blood flow, check that a baby is developing correctly before birth.

### Breaking Down Kidney Stones

Ultrasound waves can be used to break down kidney stones in the body so they can be removed without the need for painful surgery.

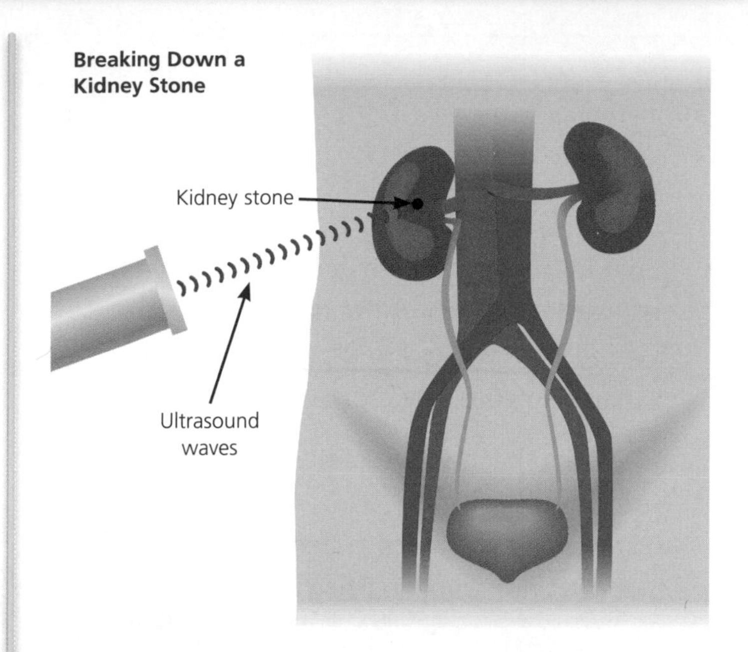


HT Ultrasonic waves cause the kidney stones to vibrate, making them break up and disperse. They are then passed out of the body in urine.

### Body Scans

Ultrasound waves are used to build up a picture of the organs in the body, including the heart, lungs and liver. They can also be used to detect gallstones and tumours. They are used for pre-natal scanning because there is no risk to either the mother or the baby.

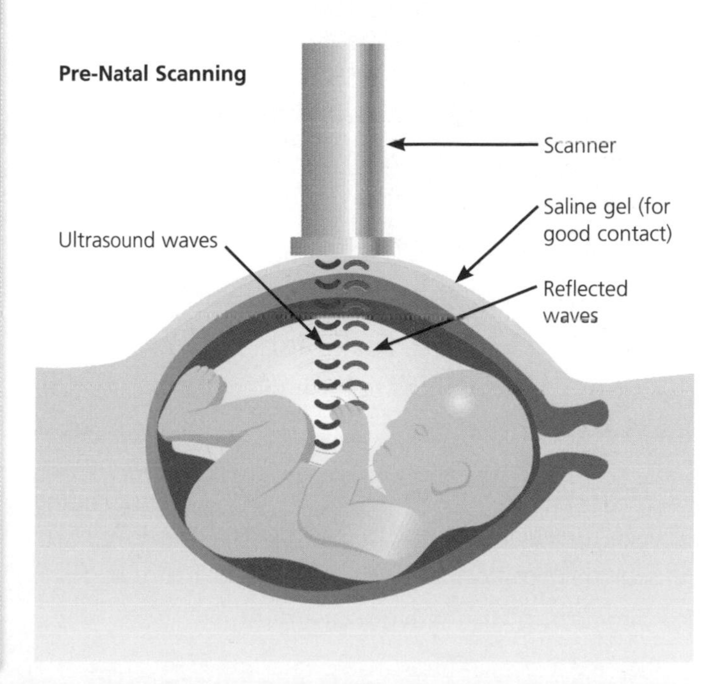

## More on Ultrasound

Ultrasound waves are partly reflected at a boundary as they pass from one medium (or substance) into another. The time taken for these reflections to be detected can be used to calculate the depth of the reflecting surface. The reflected waves are usually processed to produce a visual image on a screen.

Ultrasound has two main advantages over X-ray imaging:
- It is able to produce images of soft tissue.
- It does not damage living cells.

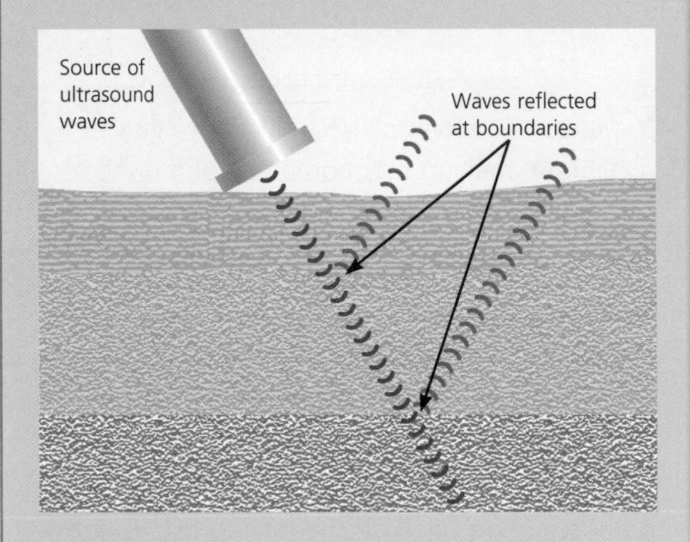


## Motion of Particles in Waves

All waves **transfer energy** from one point to another **without transferring any particles** of matter. If we allow each coil of a slinky spring to represent one particle, then we can show the movement of the particles in each wave.

**Longitudinal Waves**
Each particle moves backwards and forwards about its normal position parallel to the direction of wave movement.

**Transverse Waves**
Each particle moves up and down about its normal position at right angles (90°) to the direction of the wave movement.

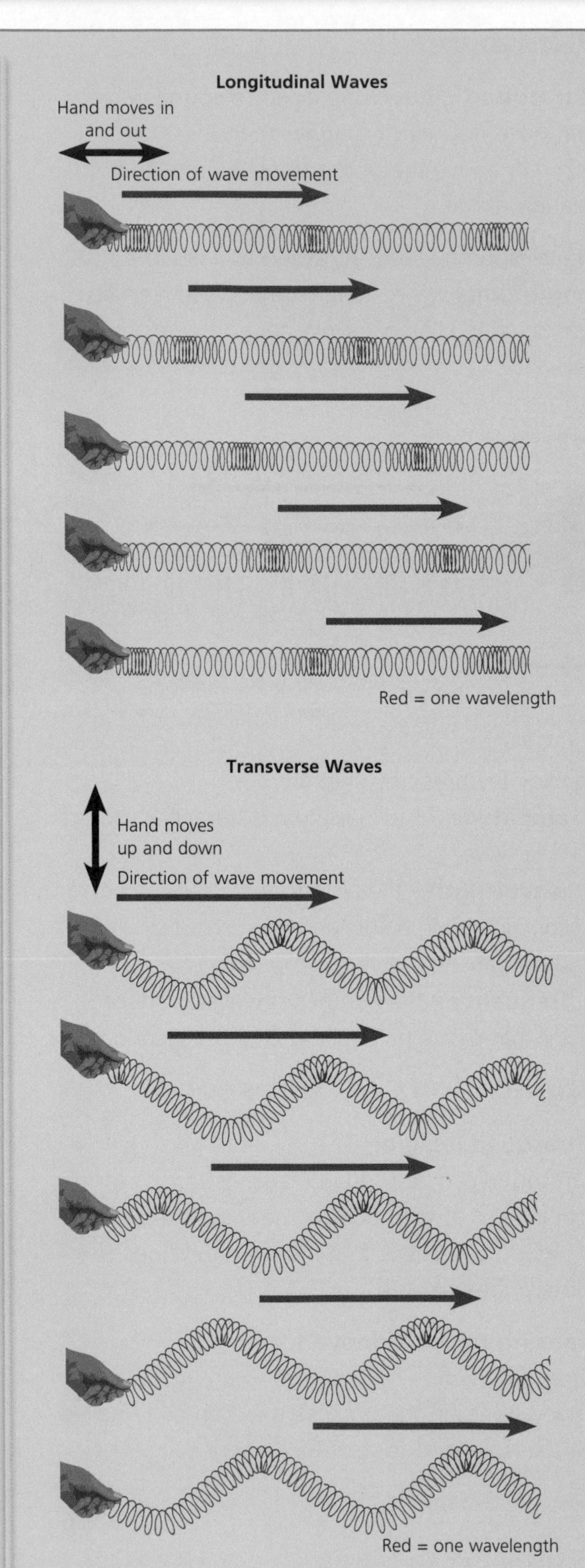

## Radioactivity

**Radioactive materials** give out **nuclear radiation** from the **nucleus** of their atoms because they are unstable and therefore decay naturally.

Radiation is measured by the number of nuclear decays emitted per second. The radioactivity of any radioactive substance decreases with time.

During decay, radiation can be given out in the form of **alpha**, **beta** and **gamma** radiation:

- An alpha particle is like a helium nucleus, i.e. two protons and two neutrons.
- A beta particle is a fast-moving electron.
- A gamma ray is an electromagnetic wave.

When any of this radiation hits an atom it may ionise it. This means it strips off an electron leaving the atom with a positive charge – an **ion**.

HT

## Alpha Emission

During alpha emission, an atom decays by ejecting an alpha particle (made up of two protons and two neutrons) from the nucleus. A new atom is formed.

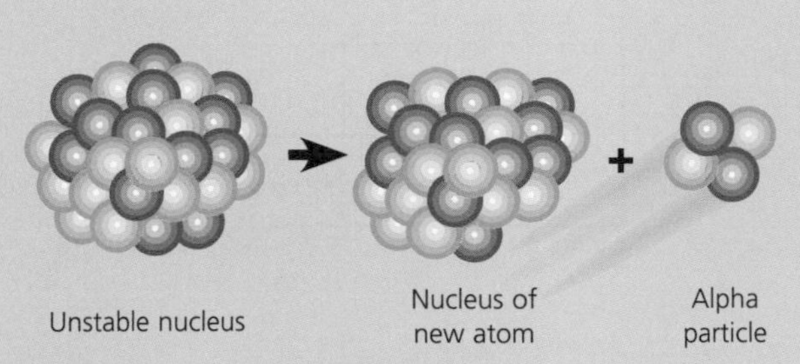


The nucleus of the new atom differs from the original one in the following ways:

- It is a different element.
- It has 2 fewer protons and 2 fewer neutrons.
- The atomic number has decreased by 2.
- The mass number has decreased by 4.

For example, the alpha decay of radium-226 into radon-222 is shown by the following equation:

$$^{226}_{88}\mathrm{Ra} \rightarrow \, ^{222}_{86}\mathrm{Rn} + \, ^{4}_{2}\alpha$$

The mass numbers (at the top) and the atomic numbers (at the bottom) balance on both sides of the equation.

Alpha particles are particularly good ionisers because they are relatively massive so they hit atoms hard enough to strip off electrons.

*N.B. Ionisation means either creating a positively charged or a negatively charged particle (ion).*

## Beta Emission

During beta emission, an atom decays by changing a neutron into a proton and an electron. The high-energy electron ejected from the nucleus is a beta particle. A new atom is formed by beta decay.

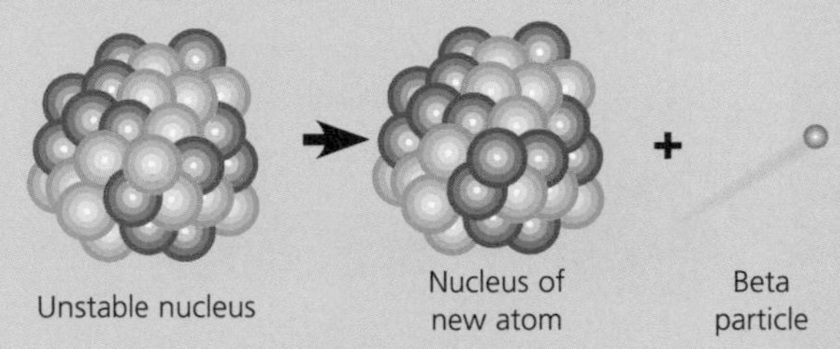


The nucleus of the new atom differs from the original one in the following ways:

- It is a different element.
- It has 1 more proton and 1 fewer neutron.
- The atomic number has increased by 1.
- The mass number remains the same.

For example, the beta decay of iodine-131 into xenon-131 is shown by the following equation:

$$^{131}_{53}\mathrm{I} \rightarrow \, ^{131}_{54}\mathrm{Xe} + \, ^{0}_{-1}\beta$$

Again, the mass numbers and the atomic numbers balance on both sides of the equation.

Beta particles are not good ionisers because they are very light and cannot easily strip an electron from an atom.

## Half-life

**Half-life** is the time it takes for half the undecayed nuclei in a radioactive substance to decay. If the substance has a very long half-life then it remains active for a very long time.

**Atoms in a Sample of Radioactive Substance**

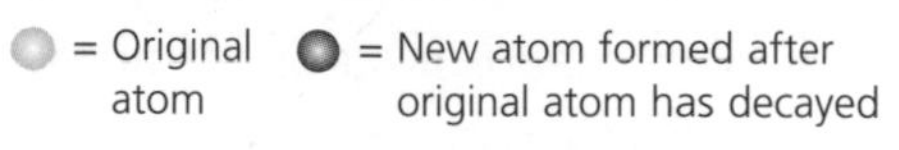


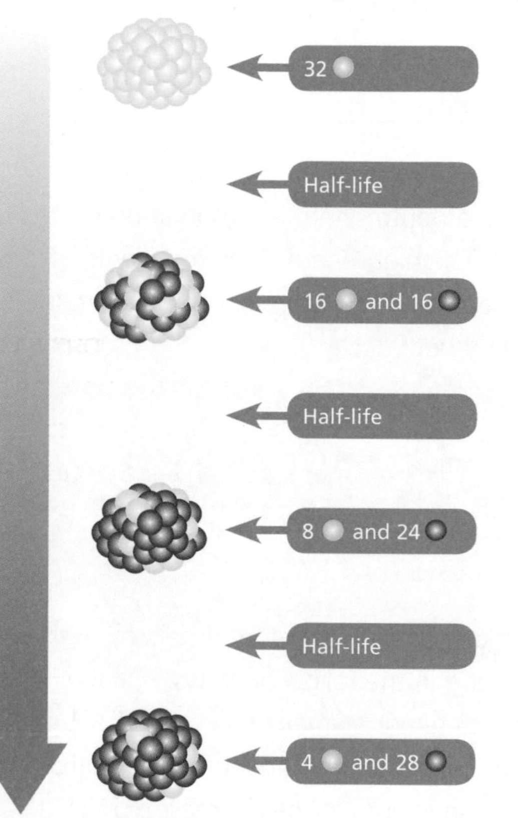


*N.B. This is a collection of atoms, not a nucleus.*

### Calculations Involving Half-life

The half-life of a substance can be calculated using a table or a graph.

**Example 1**

The table below shows the activity (measured in **becquerels, Bq**) of a radioactive substance against time.

| Time (min) | 0 | 5 | 10 | 15 | 20 | 25 | 30 |
|---|---|---|---|---|---|---|---|
| Activity (Bq) | 200 | 160 | 124 | 100 | 80 | 62 | 50 |

**a)** Calculate the half-life of the substance by using a table.

To find an average, choose three pairs of points between which the activity has halved.

| Activity | Time | Half-Life |
|---|---|---|
| 200 → 100 | 0 → 15 | 15 min |
| 160 → 80 | 5 → 20 | 15 min |
| 100 → 50 | 15 → 30 | 15 min |

The half-life is **15 minutes**.

**b)** Calculate the half-life by drawing a graph.

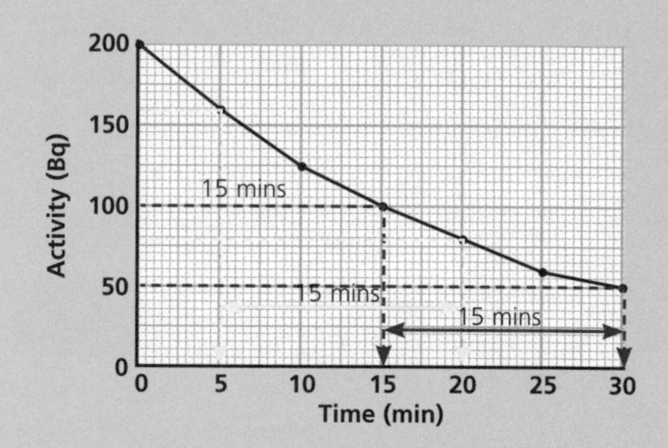


The half-life is **15 minutes**.

**Example 2**

The half-life of uranium is 4 500 000 000 years. Uranium forms lead when it decays.

A rock sample is found to contain three times as much lead as uranium. Calculate the age of the sample.

The fraction of lead present is $\frac{3}{4}$ while the fraction of uranium present is $\frac{1}{4}$. (There is three times as much lead as uranium.)

Fraction of lead + Fraction of uranium = Original amount of uranium

$$\frac{3}{4} + \frac{1}{4} = 1$$

Work out the number of decays it takes to get $\frac{1}{4}$:

$$1 \xrightarrow{\text{half-life}} \frac{1}{2} \xrightarrow{\text{half-life}} \frac{1}{4}$$ ← 2 half-lives

Age of rock = 2 × half-life

= 2 × 4 500 000 000 years

= **9 000 000 000 years**

## Background Radiation

**Background radiation** occurs **naturally** in our environment and is around us all the time. Most is released by radioactive substances in **soil** and **rocks**. The level of background radiation can vary depending on the rocks in the area. **Cosmic rays** from outer space also contribute significantly to background radiation.

> HT Not all background radiation occurs naturally. A small proportion comes from waste products and man-made sources. Industry and hospitals are both responsible for contributing to today's background radiation levels.

## Tracers

Radioisotopes (radioactive materials) are used as **tracers** in industry as well as in hospitals. They are used to find out what is happening inside objects without having to break the objects open. In industry, tracers are used to:

- track the dispersal of waste
- find leaks and blockages in underground pipes
- find the routes of underground pipes.

A radioactive material that emits gamma rays is put into the pipe. A gamma source is used because gamma can penetrate through to the surface. The progress of the material is tracked by a detector above ground:

- If there is a leak, the radioactive material will escape and will be detected at the surface.
- If there is a blockage, the radioactive material will stop flowing so it cannot be detected after this point.

## Smoke Alarms

Most smoke alarms contain americium-241, which emits alpha particles. The alpha particles cause air particles to ionise and the ions formed are attracted to oppositely charged electrodes in the alarm. This results in a current flowing in the circuit and this means the alarm is working normally.

When smoke enters the space between the two electrodes, the alpha particles are absorbed by the smoke particles and so cannot ionise the air. This causes a smaller current than normal to flow, and the alarm sounds.

**A Smoke Alarm**

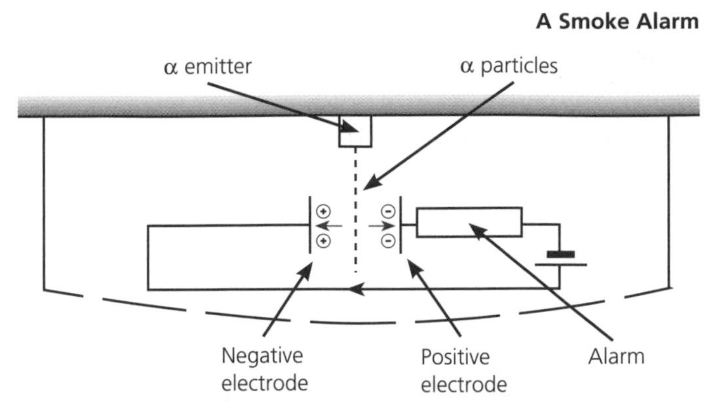


## Carbon Dating

A small amount of the carbon in our atmosphere and in the bodies of animals and plants is radioactive carbon-14. Measurements from radioactive carbon can be used to date old materials that were once alive.

> HT The amount of radioactive carbon-14 in the atmosphere has not changed for thousands of years. When an object dies it no longer exchanges gases with the air like living matter does. Therefore, as the carbon-14 in the dead object decays, the radioactivity of the sample decreases. This means that the dead object has a different radioactivity level to living matter. The ratio of these two activities can be used to find a fairly accurate age within known limits.

## Dating Rocks

Rocks were never living things so we cannot find the age of rocks using carbon dating. Instead we look at the amount of naturally occurring radioactive uranium and lead in rocks.

We look at the ratio of uranium to lead in a rock to find out its age. Radioactive uranium has a half life of 4 500 000 000 years. The ratio tells us how long the rock has existed.

## Radiation

Although some materials are radioactive naturally, it is possible to make a material radioactive by putting it into a nuclear reactor and allowing it to absorb extra neutrons. This makes its nucleus **unstable** so it emits radiation.

**X-rays** and **gamma rays** are both electromagnetic waves with similar wavelengths, but they are produced in different ways. X-rays and some nuclear radiation (i.e. **gamma** and **beta radiation**) can be used in medicine. X-rays pass easily through soft tissue and less easily through bone, which produces a shadow image of the insides of the body. The person who takes X-rays and uses radiation is called a **radiographer**.

A radiographer has to take special precautions (e.g. standing behind a lead screen) when X-raying a patient so that he/she does not receive a dangerous accumulated dose of radiation.

Gamma rays damage cells, so they can be used to treat cancer by killing cancerous cells. Beta and gamma rays pass through skin, so they can be used as medical tracers (tracking the tracer's progress through a patient's system) to help diagnose problems such as blood clots. Gamma rays can also be used to **sterilise** medical equipment because they kill germs and bacteria.

It is important that radioactive tracers do not remain in the body for long so that patients do not irradiate other people. This means that tracers must have a suitable half-life.

HT X-rays are made by firing high-speed electrons at metal targets. (The electrons lose energy very quickly.) X-rays are easier to control than gamma rays.

Gamma rays are frequently given out after alpha or beta decay. The nucleus sometimes contains surplus energy after emitting alpha or beta particles. It emits this extra energy as gamma radiation, which is very high frequency electromagnetic radiation. Unlike alpha or beta decay, gamma has no effect on the structure of the nucleus.

## HT Uses of Gamma Rays

### Treating Cancer

A wide beam of gamma rays is focused on the tumour. The beam is rotated around the outside of the body with the tumour at the centre of rotation. This concentrates the gamma rays on the tumour, but minimises damage to the rest of the body. Gamma radiation treatment destroys cancer cells without surgery. But, it may damage other (healthy) cells and cause sickness.

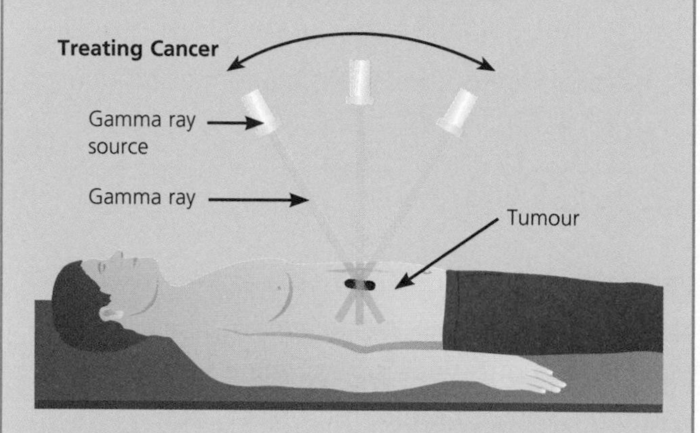


### Tracers

A tracer is a small amount of a radioactive material which is put into a patient so that its progress through the body can be followed using a radiation detector outside the body. The radioactive material must emit either gamma or beta radiation – both of these are capable of passing out of the body to be detected. It must have a short half-life so that the patient does not remain radioactive for long.

The tracer can be swallowed or injected and it is then given time to spread through the patient's body. For example, the thyroid gland is an important organ. Iodine is absorbed in the thyroid gland, so a patient is given a radioactive substance that contains iodine-131. A radiation detector can then be used to follow its progress and find out how well the thyroid gland is working by measuring the amount of iodine it absorbs.

## Fission

**Nuclear Power Station**

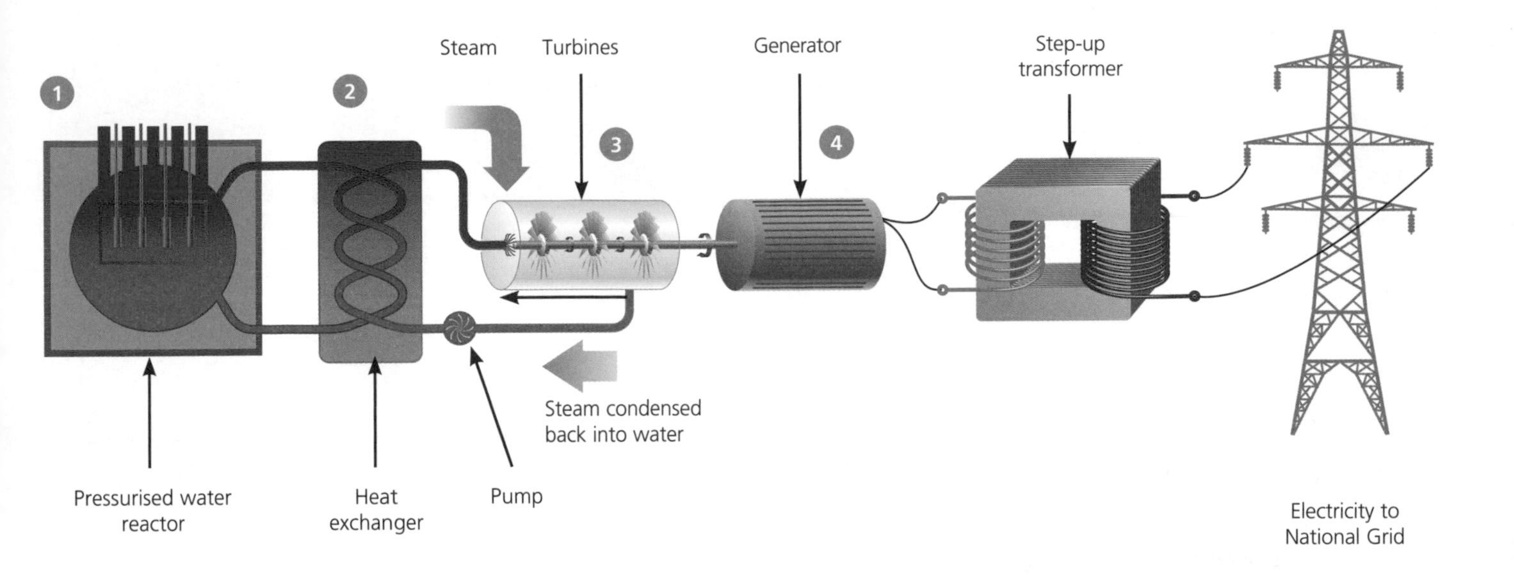


**Conventional power** stations use **fossil fuels**, i.e. coal, oil and gas, as an energy source to generate electricity. **Nuclear power stations** use **uranium** as the energy source.

1. A nuclear reaction takes place to produce heat.
2. The heat creates steam from water.
3. The steam turns a turbine.
4. The turbine turns a generator, to produce electricity.

## The Nuclear Reaction – Fission

Uranium is used to produce heat energy in a nuclear reactor. The reaction is called **nuclear fission** and this happens in a **chain reaction.** The chain reaction is carefully controlled in a power station.

Absorbing a single neutron is enough to cause a uranium nucleus to split, releasing heat energy and more neutrons. These neutrons cause more uranium nuclei to split and so the chain reaction continues.

Nuclear fission produces radioactive waste. A **nuclear bomb** is a chain reaction that has gone **out of control**, resulting in the release of one powerful burst of energy.

## HT Nuclear Fission

Nuclear fission is the process used in nuclear reactors to produce energy to make electricity.

**On a Small Scale**
Bombarding a uranium atom with a neutron causes the nucleus to split and energy is released as a result.

**On a Large Scale**
When a neutron collides with a very large nucleus (e.g. uranium), the nucleus splits up into two smaller nuclei (e.g. barium and krypton). This releases more than one neutron, which is capable of causing further fission. This is a chain reaction, so it carries on and on and on.

Scientists stop nuclear reactions getting out of control by putting control rods in the reactor. The control rods absorb some of the neutrons (preventing further fissions). The control rods can be lowered or raised to control the number of neutrons available for fission, which allows the process to keep operating safely.

## Nuclear Fission (cont)

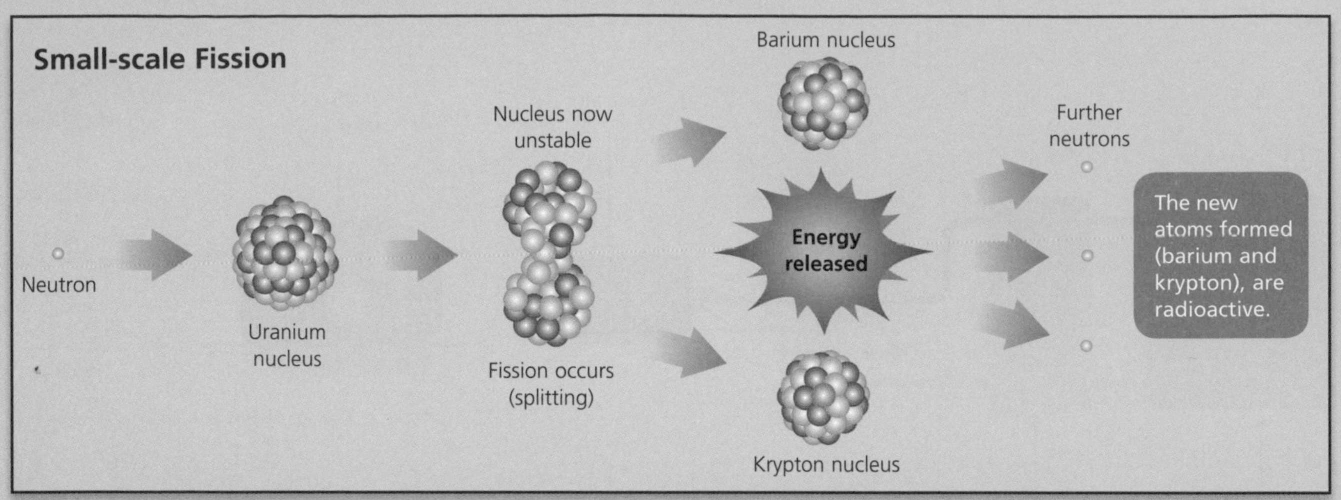


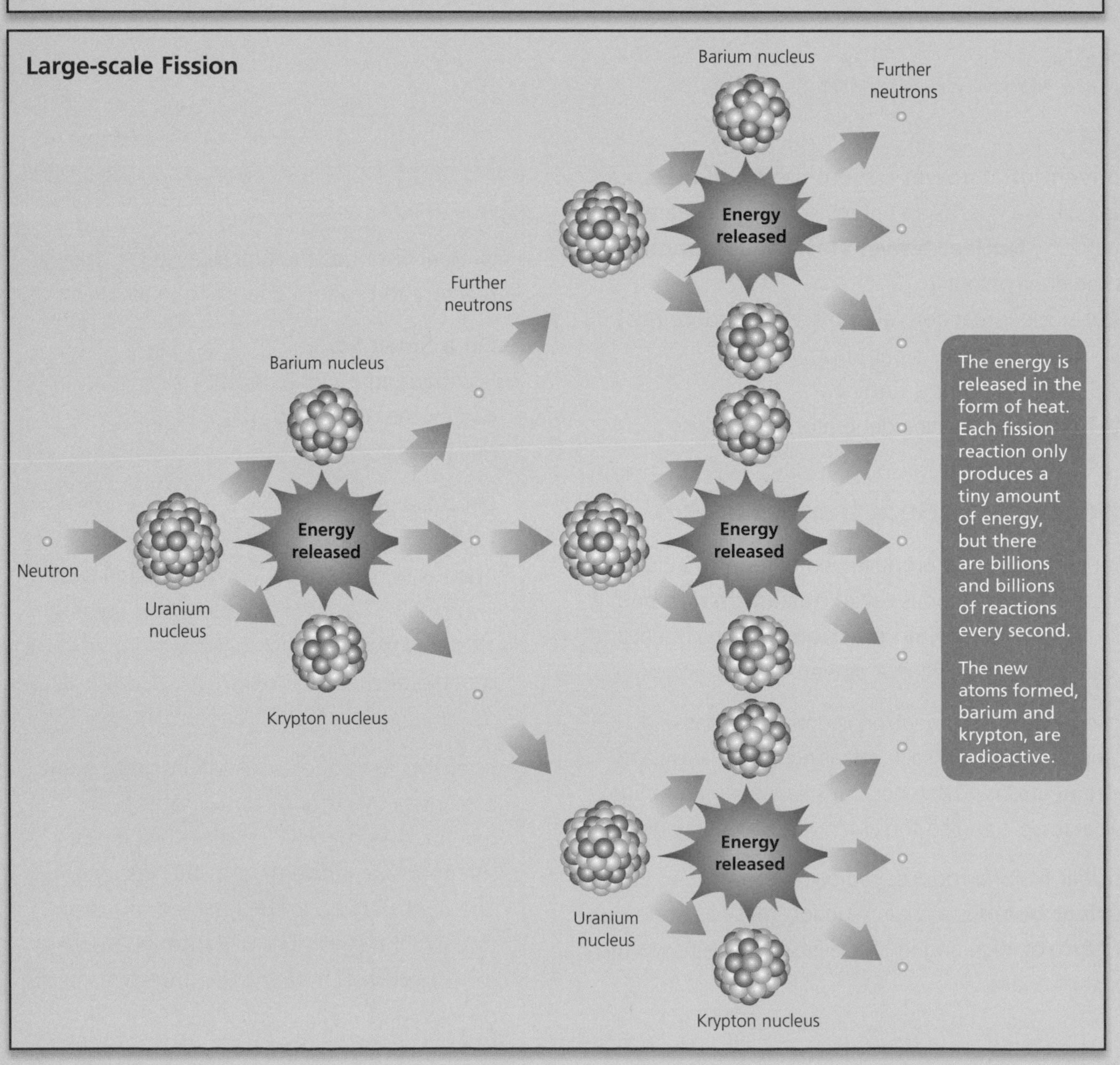

## A Nuclear Reaction – Fusion

Nuclear fusion is a different way to release energy. It is the way that stars release energy and it is what happens in a fusion bomb (also called an H bomb).

**Fusion** means two small atomic nuclei joining together. This can only happen at very high temperatures. So, for example, two hydrogen nuclei can fuse to form helium. This reaction releases a lot of energy.

## Fusion – Energy for the Future?

Nuclear fusion would be ideal for generating electricity in power stations because it produces less pollution than a fission reactor. However, it is very difficult because it requires extremely high temperatures to get it to work and the components of the power station can melt if they get this hot causing safety problems. Scientists are working to solve this problem at the moment. As it is such a big problem many countries are working together to share costs and expertise.

## Cold Fusion?

Two scientists thought they had managed to get fusion to work at room temperature. Many other scientists have tested their claims by trying to get the same experiment to work in their labs (this is how all science is tested). So far no one else has managed to get this to work so the original claims are still not proven and the idea of cold fusion is not accepted.

HT The simple difference between fission and fusion is that:

- fission is splitting an atom
- fusion is joining atoms to form larger ones.

One fusion reaction is to use two isotopes of hydrogen ($^{1}_{1}H$) and deuterium ($^{2}_{1}H$)

$$^{1}_{1}H + ^{2}_{1}H = ^{3}_{2}He$$

In stars the extremely high temperatures and pressures make nuclear fusion possible. In an H bomb it is difficult to obtain the high temperatures so a fission bomb is used to start the fusion reaction.

For generating power, the need to use extremely high temperatures and pressures makes this an inefficient process. This means that, at the moment, if energy is generated using nuclear fusion in a reactor then more energy would be used to create the temperatures and pressures required than would be obtained from the reaction. Scientists are working to overcome the safety and practical challenges so that we may, one day, use fusion to generate clean energy.

# Exam Practice Questions

1 This is a diagram of a plug.

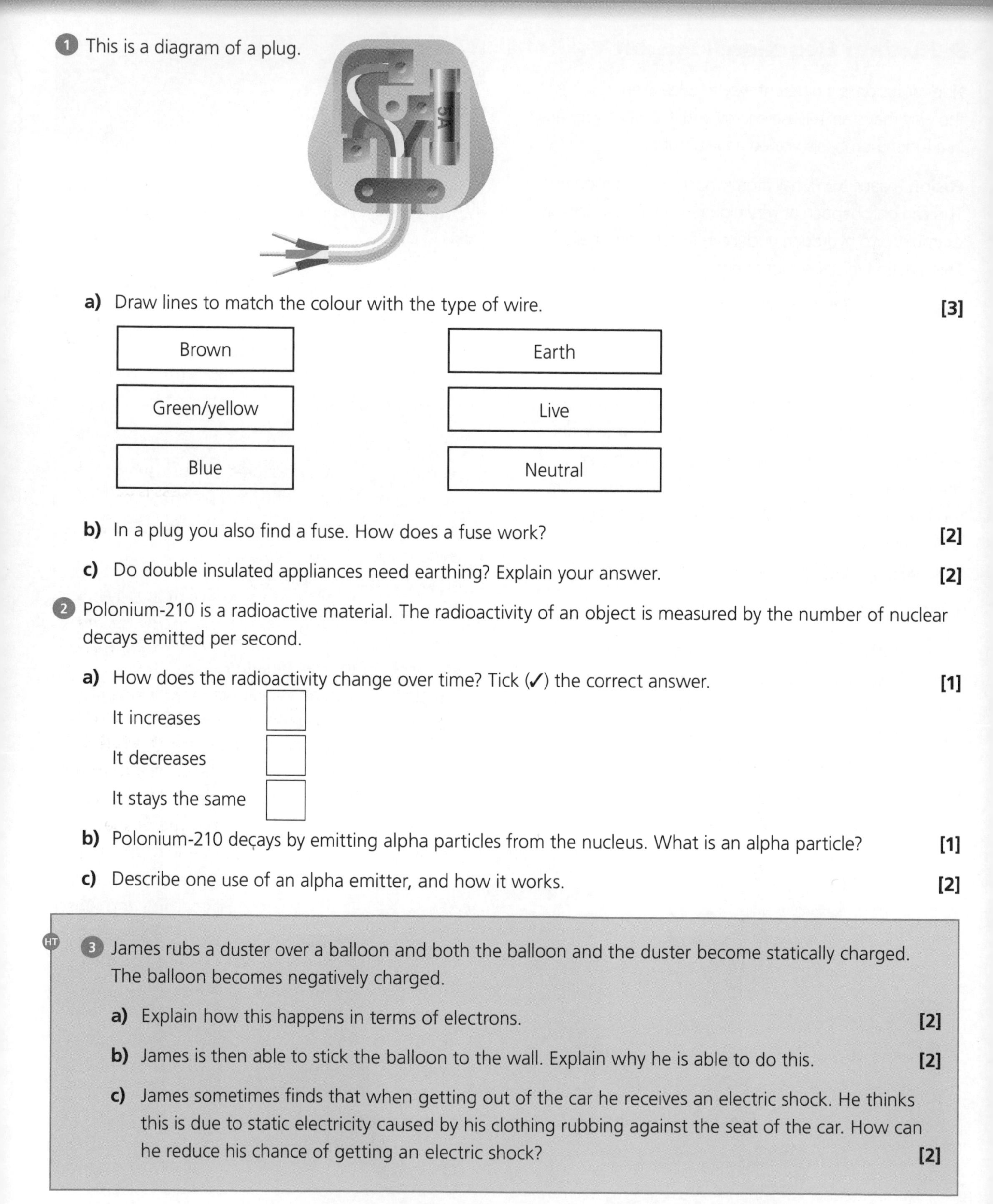


**a)** Draw lines to match the colour with the type of wire. **[3]**

| | |
|---|---|
| Brown | Earth |
| Green/yellow | Live |
| Blue | Neutral |

**b)** In a plug you also find a fuse. How does a fuse work? **[2]**

**c)** Do double insulated appliances need earthing? Explain your answer. **[2]**

2 Polonium-210 is a radioactive material. The radioactivity of an object is measured by the number of nuclear decays emitted per second.

**a)** How does the radioactivity change over time? Tick (✓) the correct answer. **[1]**

It increases ☐

It decreases ☐

It stays the same ☐

**b)** Polonium-210 decays by emitting alpha particles from the nucleus. What is an alpha particle? **[1]**

**c)** Describe one use of an alpha emitter, and how it works. **[2]**

HT

3 James rubs a duster over a balloon and both the balloon and the duster become statically charged. The balloon becomes negatively charged.

**a)** Explain how this happens in terms of electrons. **[2]**

**b)** James is then able to stick the balloon to the wall. Explain why he is able to do this. **[2]**

**c)** James sometimes finds that when getting out of the car he receives an electric shock. He thinks this is due to static electricity caused by his clothing rubbing against the seat of the car. How can he reduce his chance of getting an electric shock? **[2]**

## B3: Living and Growing

1. a)

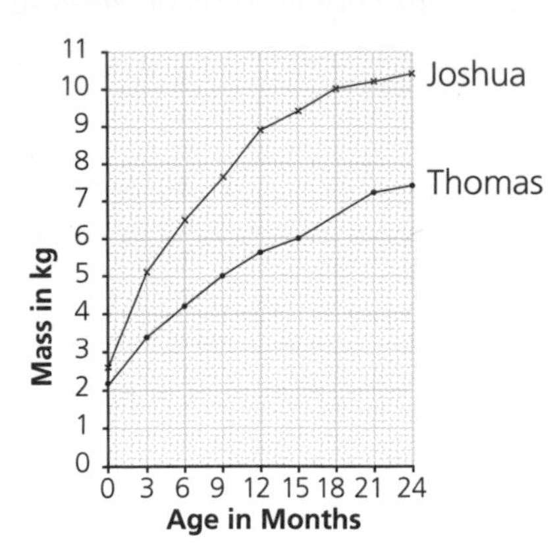


**[1 mark for correct plotting, 1 mark for curve]**

b) 6.6 kg **[+ or – 0.1kg]**

c) Between 0 and 3 months.

d) Rate of growth would increase dramatically.

e) Thomas's rate of growth would be greater than the tree's between 12 – adulthood (18 years); Tree's growth rate would continue at the same pace throughout its lifetime; Thomas's growth rate would become zero after 18-20 years / curve would level off / Thomas has no more growth during adulthood; Thomas would have slightly negative growth rate towards 50 years of age **[Any two for 2 marks]**

2. a)

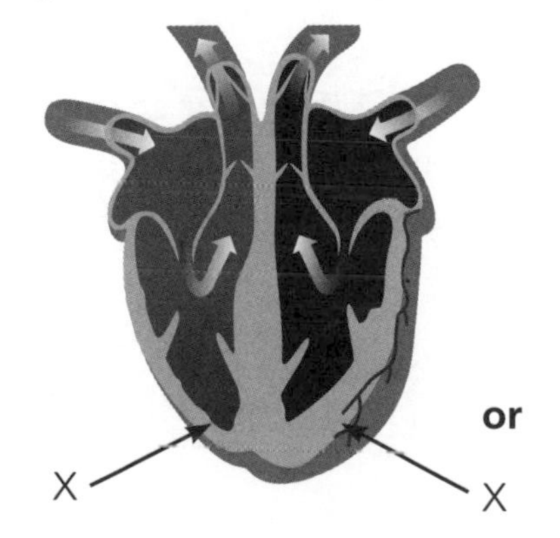


b) Layer of muscle thicker; Layer of elastic fibres thicker; Narrower lumen **[Any two for 2 marks]**

3. a) Bases / ACTG

b) Stomach pH much lower/acidic; Enzyme denatured/wrong shape of enzyme molecule prevents function.

## B4: It's a Green World

1. a) Carbon dioxide; Water.

b) Nucleic acid; Lipid; Protein; Cellulose or starch. **[Any two for 2 marks]**

c) i) Carbon dioxide

ii) Rate of bubble production would remain at a high rate (but not get any higher.)

2. a) $8 \div 43 \times 100 = 18.6\%$ **[1 mark for calculation, 1 mark for correct answer]**

b) Moving air / fan.

c) Water column in xylem would move upwards/towards the leaves.

3. a) Phosphate

b) Water uptake by osmosis; Process is passive, down a concentration gradient; Mineral ion uptake by active transport; Process is active, against a concentration gradient.

## C3 Chemical Economics

1. a) They must collide; With sufficient energy.

b) A collision must have enough energy for the particles to react; If they do not have sufficient energy they will not react.

2. **This is a model answer, which demonstrates QWC and therefore would score the full 6 marks:** The main precaution taken in a flour mill is to prevent the flour dust mixing with the air. The flour dust will burn and any type of spark will set it off so all the electrical equipment has to be made so that it does not produce a spark. Flour dust has a very large surface area and when it mixes with air there is a very large chance of collision between the particles. When the dust–air mixture is set alight there is a very fast reaction – an explosion.

3. a) 106 **b)** 46

c) i) $12 + 32 = 44$g

ii) $32 \div 12 = 2.6666$g $= 2.7$g **[1 mark for calculation, 1 mark for correct answer to 2 significant figures]**

4. a) Carbon

b) It is very hard; It has a high melting point.

5. Batch process; Because it is done on a small scale.

6. a) 26°C
   b) Energy = 100 × 4.2 × 26 = 10 920J **[1 mark for calculation, 1 mark for correct answer]**

7. Most of the bonds in graphite are strong; Carbon atoms arranged in layers but some of the bonds between layers are weak; The electrons become delocalised (able to move around).

8. $M_r$ magnesium chloride, 24 + (2 x 35.5) = 95
   24g magnesium makes 95g magnesium chloride **[1 mark]**
   9.6g magnesium makes $\frac{95 \times 9.6}{24}$g = 38g **[1 mark]**

9. a) B b) C
   c) The steeper slope is the faster reaction; Calculate the gradient.

## C4 The Periodic Table

1. a) Atoms of the same element; With different numbers of neutrons.
   b) 17 protons; 18 neutrons; 17 electrons
   c) 2.8.4

2. a) Remove 1 electron from the atom.
   b) Add 2 electrons to the atom.

3. a) Chlorine
   b) Sodium iodide + chlorine ➔ sodium chloride + iodine

4. a) They are shiny.
   b) Non-toxic; Hard; Unreactive. **[Any one for 1 mark]**
   c) It is harder; It is more lustrous. **[Any one for 1 mark]**

5. a) sodium iodide + chlorine ➔ sodium chloride + iodine
   b) It shows that chlorine is more reactive than iodine.

6. a) **Benefit** – No energy lost during power transmission; Super fast circuits; Powerful electromagnets. **[Any one for 1 mark]**
   **Problem** – They only work at very low temperatures; Expensive. **[1 mark]**
   b) Because it is a material that conducts electricity with very little or no resistance so it would create very fast electronic circuits.

7. a) Calcium
   b) $Fe^{2+} + 2OH^- \rightarrow Fe(OH)_2$ **[1 mark for correct formula, 1 mark for correct balancing]**

8. It has equal numbers of protons and electrons.

9. **This is a model answer, which demonstrates QWC and therefore would score the full 6 marks:** There are strong forces holding the ions together in solid sodium chloride and a lot of energy is needed to break them apart. When sodium chloride is molten, the charged ions can move and carry the electrical charge. In methane there are no charged particles and all the electrons are involved in bonding so there is nothing to carry an electrical charge. The forces between methane molecules are weak and so only a small amount of energy is needed to separate the molecules.

10. a)

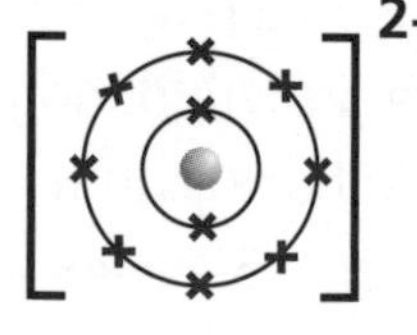


$Mg^{2+}$ ion $[2.8]^{2+}$

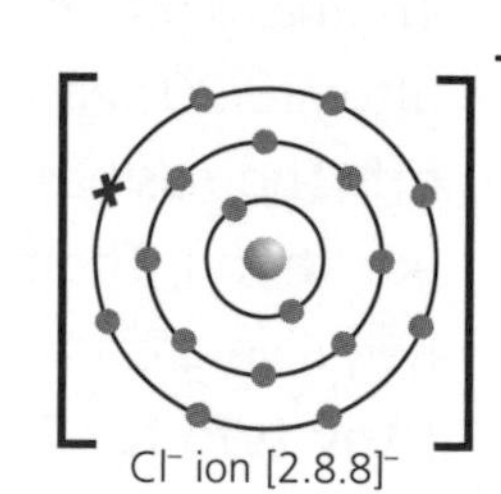


$Cl^-$ ion $[2.8.8]^-$

   b) $MgCl_2$
   c) $Br_2 + 2e^- \rightarrow 2Br^-$ **[1 mark for correct reactants, 1 mark for correct products]**. The bromine gains electrons. **[1 mark]**

11.

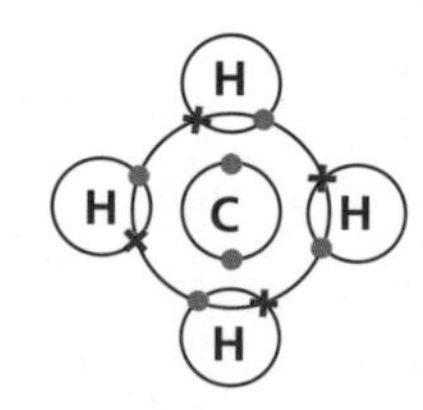


**[1 mark for one shared pair of electrons, 1 mark for the rest of the diagram]**

12. 8 protons, 10 neutrons, 8 electrons

13. **a) Sodium atom** **b) Sodium ion**

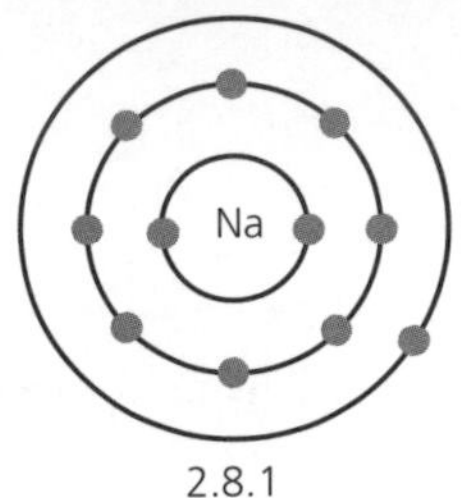


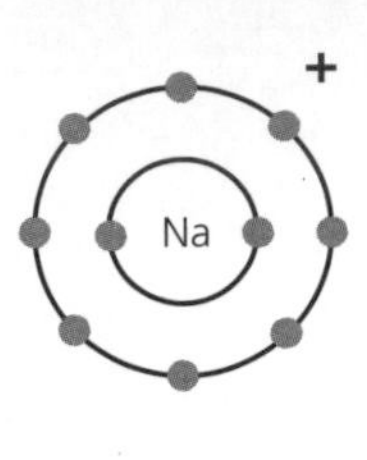


2.8.1 **[1 mark]** 2.8 **[1 mark]**

## P3: Forces for Transport

1. **a) i)** A to B – He is travelling at a steady speed away from home
   **ii)** B to C – He is stationary
   **iii)** C to D – He is travelling at a steady speed towards home
   **b)** Faster on the way there than on the way back (steeper slope).
   **c)** Away from home = $\frac{300}{50}$ = 6 m/s **[1 mark]**
   Towards home = $\frac{300}{100}$ = 3 m/s **[1 mark]**
   **d)**

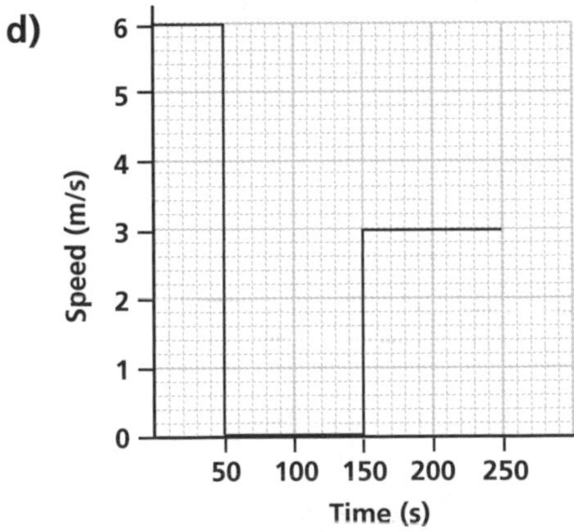


**[2 marks for correct lines, 1 mark for correct labelling of axes]**

2. **a)** 35 + 45 = 80m
   **b) Thinking distance:** The speed of the car **[1 mark]**; The reaction time of the driver e.g. the driver has been drinking alcohol. **[1 mark]**
   **Braking distance:** The weight of the car; The speed of the car; The road conditions. **[Any two for 2 marks]**.
   **c)** Absorb energy during a crash by crumpling up **[1 mark]**; This dissipates the energy more slowly, reducing the momentum of the crash **[1 mark]**; Reducing injuries to the passengers. **[1 mark]**

3. **a)** change in momentum = $m_1v_1 - m_2v_2$ = (1300 × 22.3) – (1300 × 13.4) = 28990 – 17420 = 11570kgm/s **[1 mark for calculation, 1 mark for correct answer]**
   **b) i)** change in momentum = $m_1v_1 - m_2v_2$ = (1300 × 13.4) – (1300 × 0) = 17420 – 0 = 17420kgm/s **[1 mark for calculation, 1 mark for correct answer]**
   **ii)** Force = change in momentum ÷ time = 17420 ÷ 2.2 = 7918.2N **[2 marks for calculation, 1 mark for correct answer]**

4. **a)** GPE = $mgh$
   = 80 × 10 × 30
   = 24000J **[1 mark for calculation, 1 mark for correct answer]**
   **b)** KE = GPE
   KE = 24000
   24000 = $\frac{1}{2}mv^2$
   $v$ = 24.4948....
   $v$ = 24.5m/s
   **[1 mark for calculation, 1 mark for correct answer]**
   **c)** Acceleration = $\frac{\text{change in speed}}{\text{time}}$
   $= \frac{-24.5}{2} = -12.25\text{m/s}^2$
   **[1 mark for calculation, 1 mark for correct answer]**

## P4: Radiation for Life

1. **a)** Brown – Live
   Green/yellow – Earth
   Blue – Neutral
   **b)** If the current becomes too large the fuse melts **[1 mark]**, breaking the circuit. **[1 mark]**
   **c)** No **[1 mark]**; They do not need earthing because the case is a non conductor and cannot become live. **[1 mark]**

2. **a)** It decreases.
   **b)** A helium nucleus or two protons + two neutrons.
   **c)** In a smoke detector **[1 mark]**; Smoke particles hit by alpha particles; Less ionisation of air particles, causing the current to be reduced and the alarm to sound. **[Any one for 1 mark]**

3. a) Rubbing the two insulating materials together allows electrons to transfer between objects. **[1 mark]**
The duster becomes positively charged due to lack of electrons and the balloon becomes negatively charged due to an excess of electrons. **[1 mark]**
b) The balloon repels the electrons on the atoms on the surface of the wall; **[1 mark]** This gives the wall a slight positive charge at the surface and the balloon is then attracted to the wall as opposite charges attract. **[1 mark]**
c) Using shoes with insulating soles; **[1 mark]** Not touching the metal parts of the car until the charge on his body has leaked away. **[1 mark]**

**Acceleration** – rate of increase of speed (calculated as speed change divided by time taken and measured in $m/s^2$) or change in direction of an object.

**Aerobic respiration** – energy release requiring oxygen.

**Air resistance** – the frictional force that acts on a moving object.

**Alpha radiation** – nuclear radiation particle made up of 2 protons and 2 neutrons.

**Amplitude** – the maximum disturbance of a wave from a central position.

**Anaerobic respiration** – energy release not requiring oxygen.

**Artery** – large blood vessel with narrow lumen and thick elastic walls.

**Atom** – the smallest part of an element that can enter into chemical reactions.

**Atom economy** – a measure of how much of the reactants are converted into useful product.

**Atomic number** – the number of protons in an atom.

**Attraction** – the drawing together of materials with different charges.

**Background radiation** – radiation from soil, rocks and other things in the environment.

**Base** – compound in DNA based on nitrogen that forms the 'rungs' of the DNA 'ladder'. Four types: A, C, T, G.

**Batch process** – a process in which chemicals are put into a container, the reaction takes place, and the products are removed before a new reaction is started.

**Beta radiation** – nuclear radiation made up of a fast moving electron.

**Biological control** – use of a natural predator for controlling pest organisms.

**Braking distance** – the distance that a car travels during braking to a stop.

**Capillary** – very small blood vessels forming a network around all body tissues.

**Catalyst** – a substance that is used to speed up a chemical reaction without being chemically changed at the end of the reaction.

**Cellulose** – complex carbohydrate molecule found in plant cell walls.

**Chain reaction** – in a nuclear power station the chain reaction allows the nuclear fusion to continue.

**Charge** – an electron carries a negative charge.

**Circuit** – a continuous link of conductors that can carry an electric current.

**Chromosome** – structure found in the nucleus of the cell carrying genetic information.

**Community** – the different populations of plant and animal species found interacting in a particular area.

**Compound** – a substance consisting of two or more atoms chemically combined together by ionic or covalent bonds.

**Concentration** – a measure of the amount of substance dissolved in a solution.

**Continuous process** – a process that does not stop; reactants are fed in at one end and products are removed at the other end at the same time.

**Covalent bond** – a bond between two atoms formed by sharing a pair of electrons.

**Decomposer** – organism that breaks down dead organic matter.

**Differentiation** – process of cell specialisation.

**Diffusion** – passive movement of molecules in a liquid or gas from a high to low concentration.

**Diploid** – a full chromosome set as found in most cells of the body.

**Discharge** – to remove charge.

**Distance–time graph** – a graph showing distance travelled against time taken; the gradient of the line represents speed.

**DNA** – large molecule carrying a genetic code that makes up chromosomes.

**Drag** – resistive or frictional force on moving objects.

**Earthing** – connection between the metal casing of a device and the Earth.

**Ecosystem** – the organisms and their influence on physical factors in a prescribed area.

**Electron** – a negatively charged particle that orbits the nucleus of an atom; a charged particle that flows through wires as an electric current.

**Element** – a substance that consists of only one type of atom.

**Endothermic** – a reaction in which energy is taken in.

**Enzymes** – molecules acting as biological catalysts.

**Exothermic** – a reaction in which energy is given out.

**Fatigue** – state of muscles after lactic acid has built up during anaerobic respiration.

**Fermenter** – industrial vessel used to synthesise biotechnological products; also used in the brewing industry.

**Fertilisation** – joining of egg/ovum and sperm to form a zygote.

**Force** – a push or pull acting on an object; measured in newtons (N).

**Frequency** – of AC – the number of cycles completed each second; of waves – the number of waves produced (or that pass a particular point) in one second.

**Friction** – the resistive force between two surfaces as they move over each other.

**Fungicide** – agricultural chemical used to control fungal growth in crops.

# Glossary

**Fuse** – thin metal wire designed to melt to break a circuit when it is overloaded.

**Gamete** – sex cell; sperm or egg/ovum.

**Gamma radiation** – nuclear radiation that is high frequency electromagnetic waves.

**Gene** – short section of a chromosome that codes for a protein molecule.

**Gene therapy** – treatment for inserting genes into an individual's cells and biological tissues to treat disease, such as cancer.

**Genetic engineering** – human manipulation of the genetic code in organisms.

**Genetic modification** – human manipulation of an organism's genetic material, often connected with raising yields in crop plants.

**Glucose** – basic sugar molecule used for release of energy in organisms.

**Gravitational potential energy** – the energy an object has because of its mass and height above Earth.

**Group** – a vertical column of elements in the Periodic Table.

**Habitat** – part of the physical environment where an organism lives.

**Halide** – a compound containing a metal and a halogen.

**Halogens** – elements in Group 7 of the periodic table.

**Haploid** – a half-chromosome set in gametes, as produced by meiosis.

**Herbicide** – agricultural chemical used to control growth of weeds.

**Humidity** – the water vapour content of air.

***In vitro* fertilisation (IVF)** – fertility treatment where fertilisation occurs outside of the female's body.

**Insecticide** – agricultural chemical used to control insect pests.

**Insoluble** – a substance that is unable to dissolve in a solvent.

**Insulation** – material containing air pockets that reduces heat loss by conduction.

**Intensive farming** – modern farming methods characterised by high yields and application of man-made chemicals.

**Ion** – a positively or negatively charged particle formed when an atom gains or loses one or more electron(s).

**Ionic bond** – the bond formed when electrons are transferred between a metal atom and a non-metal atom, creating charged ions that are held together by forces of attraction.

**Isotopes** – atoms of the same element that contain different numbers of neutrons.

**Kinetic energy (KE)** – the energy possessed by a body because of its movement.

**Lactic acid** – waste product formed in muscles by anaerobic respiration.

**Limiting reactant** – the reactant that gets used up first in a reaction.

**Longitudinal wave** – an energy-carrying wave where the particles of the medium move in the direction of energy transfer.

**Mass** – the quantity of matter in an object.

**Mass number** – the total number of protons and neutrons in an atom.

**Meiosis** – cell division in reproductive tissue that results in production of gametes with a half-chromosome set.

**Meristem** – a region of actively dividing cells in plants, e.g. root or shoot tip.

**Mitosis** – cell division producing identical cells during growth, repair and asexual reproduction.

**Molecule** – two or more atoms bonded together.

**Momentum** – a measure of the state of motion of an object; product of its mass and velocity.

**Mutation** – change in the base sequence of DNA resulting in a different protein being formed.

**Neutron** – uncharged particle in the nucleus of an atom.

**Nanochemistry** – the study of materials that have a very small size, in the order of 1–100nm; one nanometre is one billionth of a metre and can be written as 1nm or $1 \times 10^{-9}$m.

**Neutron** – a particle found in the nucleus of atoms; it has no charge; relative mass 1.

**Nuclear fission** – the splitting of atomic nuclei giving out energy.

**Nuclear fuel** – non-renewable fuel that generates heat from fission reactions e.g. uranium or plutonium.

**Nuclear fusion** – two atoms joining together giving out energy.

**Nucleus** – the core of an atom, made up of protons and neutrons (except hydrogen, which contains a single proton).

**Organic farming** – farming method avoiding the use of chemical fertilisers and pesticides.

**Osmosis** – special diffusion of water from high to low water potential (i.e. low solute to high solute concentration).

**Oxidation** – a reaction involving the gain of oxygen or the loss of electrons.

**Partially permeable membrane** – membrane that allows only certain molecules or atoms through, graded by size.

**Period** – a horizontal row of elements in the Periodic Table.

**Periodic Table** – a list of the elements arranged to show the trends and patterns in their properties.

**Phloem** – cell tubes in plants that transport carbohydrates.

**Pollutant** – a chemical produced by human activity that can harm the environment and organisms.

**Pooter** – animal sampling apparatus consisting of a small vessel and two tubes; used to 'suck up' flying insects.

**Population** – number of individuals of a single species found in a given area.

**Power** – the rate of doing work; measured in watts (W).

**Precipitate** – an insoluble solid formed during a reaction involving solutions.

**Precipitation** – the formation of an insoluble solid (a precipitate) when two solutions containing ions are mixed together.

**Product** – a substance produced in a reaction.

**Proton** – a positively charged particle found in the nucleus of an atom; relative mass 1.

**Quadrat** – square frame used for sampling plants.

**Radioactive** – substance that emits radiation.

**Reactant** – a starting material in a reaction.

**Reduction** – a reaction involving the loss of oxygen or the gain of electrons.

**Relative speed** – the speed of an object, relative to another object that is treated as being at rest.

**Repulsion** – the pushing away of materials that have the same charge.

**Resistance** – how hard it is to get a current through a component at a particular potential difference; measured in ohms (Ω).

**Respiration** – energy release in cells.

**Rheostat** – variable resistor used to control electric current.

**Salt** – the product of a chemical reaction between a base and an acid.

**Solvent** – a liquid that can dissolve another substance to produce a solution.

**Speed** – the rate at which an object moves in m/s – a scalar quantity.

**Speed–time graph** – a graph showing speed against time; the gradient of the line represents acceleration; the area under the line represents distance.

**Starch** – complex carbohydrate molecule built up from individual sugar 'units'.

**Static electricity** – electricity that is produced by friction and does not flow.

**Stopping distance** – calculated as thinking distance + braking distance.

**Thinking distance** – distance that a car travels while the driver reacts and starts to brake.

**Tracer** – a radioactive substance that can be followed and detected.

**Trajectory** – the path followed by a projectile (moving body).

**Translocation** – movement of carbohydrate in solution through a plant.

**Transpiration** – flow of water through a plant, particularly loss of water from leaves.

**Transverse wave** – a wave in which the vibrations are at 90° to the direction of energy transfer.

**Ultrasound** – sound waves with a frequency above 20,000 Hz.

**Vein** – large blood vessel with relatively thin walls and valves.

**Velocity** – an object's rate of displacement (change of distance) in a particular direction; a measure of displacement divided by time – a vector quantity.

**Weight** – the force of an object on the Earth due to gravity pulling the object towards the centre of the Earth.

**Work** – energy used moving a force through a distance; measured in joules(J).

**Xylem** – dead cell tubes in a plant that transport water and minerals.

**Yield** – the amount of product obtained, e.g. from a crop or a chemical reaction.

**Zygote** – a fertilised egg cell.

**Half-life** – the time taken for half the atoms in radioactive material to decay.

**Lumen** – hollow centre of a tube or vessel as found in blood vessels.

**Lysis** – splitting (either cells or molecules).

**Metabolic rate** – the speed with which chemical reactions in the body occur, particularly respiration reactions.

**Oxygen debt** – state of muscles representing the need to 'pay back' oxygen after anaerobic respiration has been used. Replenishing of oxygen removes lactic acid.

**Plasmid** – a DNA molecule that is separate from the main genetic material found in bacteria; often circular in appearance.

**Plasmolysed** – state of plant cells when they have lost water and the cell membrane has pulled away from the cell wall.

**Saprophyte** – organism that feeds through absorbing nutrients in solution (usually in dead/decaying material).

**Terminal speed** – a steady falling speed, when the weight of an object is equal and opposite to its air resistance.

# Periodic Table

**Key**

| relative atomic mass<br>**atomic symbol**<br>name<br>atomic (proton) number |
|---|

| 1<br>**H**<br>hydrogen<br>1 |
|---|

| 1 | 2 | | | | | | | | | | | 3 | 4 | 5 | 6 | 7 | 0 |
|---|---|---|---|---|---|---|---|---|---|---|---|---|---|---|---|---|---|
| | | | | | | | | | | | | | | | | | 4<br>**He**<br>helium<br>2 |
| 7<br>**Li**<br>lithium<br>3 | 9<br>**Be**<br>beryllium<br>4 | | | | | | | | | | | 11<br>**B**<br>boron<br>5 | 12<br>**C**<br>carbon<br>6 | 14<br>**N**<br>nitrogen<br>7 | 16<br>**O**<br>oxygen<br>8 | 19<br>**F**<br>fluorine<br>9 | 20<br>**Ne**<br>neon<br>10 |
| 23<br>**Na**<br>sodium<br>11 | 24<br>**Mg**<br>magnesium<br>12 | | | | | | | | | | | 27<br>**Al**<br>aluminium<br>13 | 28<br>**Si**<br>silicon<br>14 | 31<br>**P**<br>phosphorus<br>15 | 32<br>**S**<br>sulfur<br>16 | 35.5<br>**Cl**<br>chlorine<br>17 | 40<br>**Ar**<br>argon<br>18 |
| 39<br>**K**<br>potassium<br>19 | 40<br>**Ca**<br>calcium<br>20 | 45<br>**Sc**<br>scandium<br>21 | 48<br>**Ti**<br>titanium<br>22 | 51<br>**V**<br>vanadium<br>23 | 52<br>**Cr**<br>chromium<br>24 | 55<br>**Mn**<br>manganese<br>25 | 56<br>**Fe**<br>iron<br>26 | 59<br>**Co**<br>cobalt<br>27 | 59<br>**Ni**<br>nickel<br>28 | 63.5<br>**Cu**<br>copper<br>29 | 65<br>**Zn**<br>zinc<br>30 | 70<br>**Ga**<br>gallium<br>31 | 73<br>**Ge**<br>germanium<br>32 | 75<br>**As**<br>arsenic<br>33 | 79<br>**Se**<br>selenium<br>34 | 80<br>**Br**<br>bromine<br>35 | 84<br>**Kr**<br>krypton<br>36 |
| 85<br>**Rb**<br>rubidium<br>37 | 88<br>**Sr**<br>strontium<br>38 | 89<br>**Y**<br>yttrium<br>39 | 91<br>**Zr**<br>zirconium<br>40 | 93<br>**Nb**<br>niobium<br>41 | 96<br>**Mo**<br>molybdenum<br>42 | [98]<br>**Tc**<br>technetium<br>43 | 101<br>**Ru**<br>ruthenium<br>44 | 103<br>**Rh**<br>rhodium<br>45 | 106<br>**Pd**<br>palladium<br>46 | 108<br>**Ag**<br>silver<br>47 | 112<br>**Cd**<br>cadmium<br>48 | 115<br>**In**<br>indium<br>49 | 119<br>**Sn**<br>tin<br>50 | 122<br>**Sb**<br>antimony<br>51 | 128<br>**Te**<br>tellurium<br>52 | 127<br>**I**<br>iodine<br>53 | 131<br>**Xe**<br>xenon<br>54 |
| 133<br>**Cs**<br>caesium<br>55 | 137<br>**Ba**<br>barium<br>56 | 139<br>**La***<br>lanthanum<br>57 | 178<br>**Hf**<br>hafnium<br>72 | 181<br>**Ta**<br>tantalum<br>73 | 184<br>**W**<br>tungsten<br>74 | 186<br>**Re**<br>rhenium<br>75 | 190<br>**Os**<br>osmium<br>76 | 192<br>**Ir**<br>iridium<br>77 | 195<br>**Pt**<br>platinum<br>78 | 197<br>**Au**<br>gold<br>79 | 201<br>**Hg**<br>mercury<br>80 | 204<br>**Tl**<br>thallium<br>81 | 207<br>**Pb**<br>lead<br>82 | 209<br>**Bi**<br>bismuth<br>83 | [209]<br>**Po**<br>polonium<br>84 | [210]<br>**At**<br>astatine<br>85 | [222]<br>**Rn**<br>radon<br>86 |
| [223]<br>**Fr**<br>francium<br>87 | [226]<br>**Ra**<br>radium<br>88 | [227]<br>**Ac***<br>actinium<br>89 | [261]<br>**Rf**<br>rutherfordium<br>104 | [262]<br>**Db**<br>dubnium<br>105 | [266]<br>**Sg**<br>seaborgium<br>106 | [264]<br>**Bh**<br>bohrium<br>107 | [277]<br>**Hs**<br>hassium<br>108 | [268]<br>**Mt**<br>meitnerium<br>109 | [271]<br>**Ds**<br>darmstadtium<br>110 | [272]<br>**Rg**<br>roentgenium<br>111 | Elements with atomic numbers 112–116 have been reported but not fully authenticated | | | | | | |

*The lanthanoids (atomic numbers 58–71) and the actinoids (atomic numbers 90–103) have been omitted.